# GRAPHING CALCULATOR AND COMPUTER GRAPHING LABORATORY MANUAL

## PRECALCULUS SERIES
### SECOND EDITION

Demana/Waits/Clemens

Franklin Demana
The Ohio State University

Bert K. Waits
The Ohio State University

Chuck Vonder Embse
Central Michigan University

Gregory D. Foley
Sam Houston State University

# GRAPHING CALCULATOR AND COMPUTER GRAPHING LABORATORY MANUAL

## PRECALCULUS SERIES
### SECOND EDITION

Demana/Waits/Clemens

**ADDISON-WESLEY PUBLISHING COMPANY**

Reading, Massachusetts • Menlo Park, California • New York
Don Mills, Ontario • Wokingham, England • Amsterdam • Bonn
Sydney • Singapore • Tokyo • Madrid • San Juan • Milan • Paris

Reproduced by Addison-Wesley from camera-ready copy supplied by the authors.

ISBN 0-201-56852-7

Reprinted with corrections, June 1992

2 3 4 5 6 7 8 9 10–MU–95949392

# ORDER FORM—VERSION 1.0 MASTER GRAPHER AND 3D GRAPHER

### DESIGNED BY
### BERT K. WAITS AND FRANKLIN DEMANA
The Ohio State University

## Five Powerful Graphing Utilities for IBM, Macintosh, and Apple II Computers

*Master Grapher 1.0 includes interactive function, polar, parametric, conic, and surface graphing utilities.* Note: *Master Grapher 1.0 now* includes *3D Grapher.*

| | | | QUANTITY | TOTAL PRICE |
|---|---|---|---|---|
| 50854 | ☐ | IBM—3.5" Disk | $32.76 | _____ $_____ |
| 50850 | ☐ | IBM—5.25" Disks | $32.76 | _____ $_____ |
| 50859 | ☐ | Macintosh | $32.76 | _____ $_____ |
| 50857 | ☐ | Apple II | $32.76 | _____ $_____ |

Note: Some discounts may apply—call 1-800-447-2226

## Master Grapher 2.0 will be available in Fall 1992

| | | | | |
|---|---|---|---|---|
| 58525 | ☐ | IBM—3.5" Disk | $32.76 | _____ $_____ |
| 54838 | ☐ | IBM—5.25" Disks | $32.76 | _____ $_____ |
| 54837 | ☐ | Macintosh | $32.76 | _____ $_____ |
| 54839 | ☐ | Apple II | $32.76 | _____ $_____ |

A site license is available.
Write: Site License Coodinator,
Addison-Wesley, 1 Jacob Way,
Reading, Massachusetts 01867
Phone: 1-617-944-3700 x2575 for details

SUBTOTAL : _____

ADD SALES TAX
WHERE APPLICABLE : _____

TOTAL ORDER : _____

## TO ORDER BY MAIL: FORM OF PAYMENT

☐ CHECK ENCLOSED—MADE PAYABLE TO ADDISON-WESLEY PUBLISHING COMPANY

☐ CREDIT CARD (CIRCLE ONE:   MASTER CARD   VISA   AMERICAN EXPRESS)

CARD NUMBER _____ Expiration Date:_____

_____
Signature

☐ SCHOOL OR UNIVERSITY PURCHASE ORDER (ENCLOSE P.O.)

SHIP TO
NAME_____ DEPARTMENT_____

SCHOOL OR COLLEGE_____ STREET_____

CITY/STATE_____ ZIP_____

MAIL THIS FORM TO:
**ADDISON-WESLEY PUBLISHING COMPANY**
**ATTN: ORDER DEPARTMENT**
**1 Jacob Way**
**Reading, MA 01867**

TO ORDER BY TELEPHONE, CALL TOLL-FREE: 1-800-447-2226

# PREFACE

This manual is designed to be used in conjunction with the Precalculus series, 2/e by Franklin Demana, Bert K. Waits, and Stan Clemens. The textbook series presents essential elements of mathematics needed to provide a solid foundation for the study of advanced mathematics and science. They present an integrated approach using current computer and calculator based graphing throughout the materials to enhance the teaching and learning of precalculus mathematics. Students are expected to have regular and frequent access to a computer and appropriate graphing software or a graphing calculator for homework outside of class and for occasional class laboratory activities as well. This laboratory manual provides an introduction for such activities.

Modern technology has evolved to the stage that it should be routinely used by mathematics students at all levels. Computer and calculator based graphing removes the need for contrived problems and opens the door for students to explore and solve realistic and interesting applications. The teaching and learning of traditional topics can be improved with the full use of technology. Computer and calculator based technology can turn the mathematics classroom into a mathematics laboratory. Technology gives rise to interactive instructional models that permit a focus on problem solving and encourage generalizations based on strong geometric evidence. The new instructional approaches possible with the use of technology make teachers and students active partners in an exciting, rewarding, enjoyable, and intensive educational experience. It is in this spirit of exploration and experimentation that the Demana-Waits-Clemens textbooks and this manual were written.

**ACKNOWLEDGMENTS**  The authors are indebted to a great many colleagues who participated in the development of this edition and previous editions of the manual. Several individuals contributed in whole, or part, various chapters of this manual. They deserve special recognition and we extend our sincere appreciation.

David New  (*The Ohio State University*)
   IBM and Mac graphing chapters
John Harvey  (*The University of Wisconsin – Madison*)
   The Sharp graphing calculator chapter
Tom Tucker  (*Colgate University*)
   The Hewlett Packard 28-S material
Sharon Sledge  (*Casio*)
   The Casio 7700 graphing calculator chapter
William C. Wickes  (*Hewlett Packard*)
   The Hewlett Packard 48SX material

We also appreciate the important contributions to this manual and previous editions of the manual made by our high school teaching partners Bruce Blackston and Dan Rohrs (*Upper Arlington High School*), Pamela Dase and Patricia Borys (*Centennial High School*), Fred Koenig (*Walnut Ridge High School*), and Ron Meyer (*Franklin Heights High School*).

We also thank Pat Oduor, Kathy Manley, Laurie Petrycki, and Sally Simpson for the expert help in producing this manual.

The existence of this manual should not be considered as an endorsement of any Apple, Casio, Hewlett Packard, IBM, Sharp, or Texas Instruments product.

Franklin Demana          Gregory D. Foley          Charles Vonder Embse          Bert K. Waits

Columbus, Ohio
June, 1991

# CONTENTS

# GRAPHING CALCULATOR AND COMPUTER GRAPHING LABORATORY MANUAL

## PRECALCULUS SERIES
### SECOND EDITION

Demana/Waits/Clemens

# CHAPTER 1

## OVERVIEW

### 1.1  Introduction

This manual is designed to be used with a series of innovative textbooks written by Franklin Demana, Bert K. Waits, and Stanley Clemens that integrate the use of interactive computer based graphing to enhance the learning of precalculus mathematics. The phrase "computer based graphing" includes the use of powerful pocket graphics computers, usually called graphing calculators, as well as graphing software on personal computers.

The manual is divided into two parts. The first part includes material on how to use the Texas Instruments, Casio, Sharp, and Hewlett Packard graphing calculators, and the second part contains material on how to use the interactive graphing software package *Master Grapher*. The *Master Grapher* package contains powerful utilities for graphing functions, conics, polar equations, parametric equations, and surfaces (functions of two variables) on IBM, Macintosh, and Apple II computers.

### 1.2  Objectives of Computer Based Graphing

A computer drawn graph is accurate and easy to obtain. Computer drawn graphs can be used as tools to solve equations and enhance the teaching, learning, and understanding of mathematics. Listed below are several important objectives for a computer graphing approach to learning mathematics.

> THROUGH THE SPEED AND POWER OF COMPUTERS,
> YOU CAN INVESTIGATE MANY EXAMPLES QUICKLY
> AND MAKE AND TEST GENERALIZATIONS
> BASED ON STRONG GRAPHICAL EVIDENCE

(1)  To study the behavior of functions and relations including conic equations, parametric equations, polar equations, and three–dimensional surfaces (functions of two variables).

(2)  To deepen understanding and intuition about a wide variety of functions and relations and to provide a foundation for the study of calculus, statistics, and higher mathematics.

(3)  To graphically determine the number of solutions to equations and systems of equations. To solve equations, systems of equations, and inequalities graphically with accuracy equal to any numerical approximation method.

(4)  To determine relative maximum and minimum values of functions graphically with accuracy equal to any numerical approximation method.

(5) To graphically investigate and determine the solution to "real world" problem situations that are normally not accessible to precalculus students.

(6) To provide geometric representations for problem situations and to analyze their connections with algebraic representations for the problem situations.

> COMPUTER GRAPHING IS A FAST AND EFFECTIVE TOOL THAT
> YOU CAN USE TO EXPLORE MATHEMATICS AND SOLVE PROBLEMS

## 1.3 Definitions

The following are definitions introduced in the textbooks that accompany this manual and are briefly reviewed here to help the reader of this manual.

(1) A **graphing utility** is a hand–held device, such as a Casio $fx$–$7000G$ graphing calculator, or a computer such as an Apple IIe with graphing software (a computer program) such as *Master Grapher*, that will quickly draw an accurate graph of a function or a relation.

(2) The **viewing rectangle** [L, R] by [B, T] (see Figure 1.1) is the rectangular portion of the coordinate plane determined by $L \leq x \leq R$ and $B \leq y \leq T$. The $[-10, 10]$ by $[-10, 10]$ viewing rectangle is called the **standard viewing rectangle**. We also use Xmin for L, Xmax for R, Ymin for B, and Ymax for T.

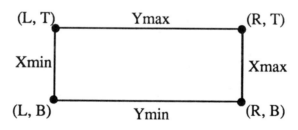

**Figure 1.1**

(3) **Zoom in** is a process of framing a small rectangular area within a given viewing rectangle, making it the new viewing rectangle, and then quickly replotting the graph in this new viewing rectangle. This feature permits the user to create a decreasing sequence of nested rectangles that "squeeze down" on a key point on a graph. Zoom-in is very useful for solving equations, inequalities, systems of equations and inequalities, and for determining maximum and minimum values of functions.

(4) **Zoom out** is a process of increasing the absolute value of the viewing rectangle parameters. It is important to be able to zoom out in *both* the horizontal and vertical directions at the same time, in *only* the horizontal direction, or in *only* the vertical direction. The zoom-out process is useful for examining limiting, end behavior of relations, and for determining "complete" graphs of relations.

(5) A **complete graph** is either the entire graph or a portion of a graph that shows all of the important behavior and features of the graph. For example, the graph of the relation $x^2 + y^2 = 16$ in $[-10, 10]$ by $[-10, 10]$ is complete because it is the entire graph of $x^2 + y^2 = 16$. The graph of $f(x) = x^3 - x + 15$ in $[-10, 10]$ by $[-10, 30]$ is a complete graph because we can see all of its local extremum values and real zeros. Of course, it is possible to create a function for which you cannot determine *one* viewing rectangle that gives a complete graph. Thus, several viewing rectangles may be needed to describe a complete graph.

(6) The **error** in using a particular point $(x, y)$ in the viewing rectangle [L, R] by [B, T] to approximate any other point $(a, b)$ in the viewing rectangle is *at most* $R - L$ for $x$ and $T - B$ for $y$. There are also better error bounds possible by using scale marks appearing in a viewing rectangle.

# Chapter 2

## Getting Started with Graphing Calculators

### 2.1 Introduction

Welcome to the frontiers of calculator technology! The purpose of Chapters 3–9 of this manual is to acquaint you with features of the major brands of graphing calculators that are useful in a graphical approach to precalculus mathematics. There are many important features of each graphing calculator that are not covered in this manual at all. Consult the *Owner's Manual* for a more complete description of your specific graphing calculator and its features.

Graphing calculators are not just calculators; they are hand-held *computers*. There are three characteristics of these versatile machines that combine to make them worthy of the name computer:

(1) **Large screen display.** The screen of a graphing calculator is large enough to display both the input (problem to be done) and output (answer). If an answer doesn't make sense or if you've made a keying or other type of error, you can go back and change (edit) any part of the input and reexecute the problem. The screen can display information in tabular form so that a progression of values can be viewed and checked for patterns.

(2) **Interactive graphics.** You can create virtually any mathematical graph: functions, relations, and geometric figures. You can overlay graphs, change views, and get a coordinate readout of specific points of interest.

(3) **On-screen programming.** The programming language of graphing calculators is simple and easy to learn. Armed with a few fundamentals you can do a great deal. Programming is ideal for repeated calculations (table building, sequences, etc.) and repeated graphs (family of curves). The programs provided in this manual are designed to turn your graphing calculator into a powerful tool for exploring and solving a wide variety of mathematical problems.

The next seven chapters provide some general information about using graphing calculators, explain how to do basic computation, discuss the basics of graphing, and offer some programming ideas to enhance your calculator's graphing capability. So get ready to explore the exciting new world of the hand-held graphics computer. The quickest way to learn about your graphing calculator is to *experiment*. Explore! Be curious! And always have your graphing calculator handy while reading the appropriate chapter of this manual.

### 2.2 Keys and Keying Sequences

In this manual, graphing calculator keys appear as boxes. So, for example, the addition key is represented by $\boxed{+}$. The first few pages of your calculator *Owner's Manual* will give a brief introduction to the keys on your specific calculator. You are urged to read it.

A **keying sequence** is always read from left to right. For example, when we write $\boxed{7}$ $\boxed{\div}$ $\boxed{4}$, you should press the three keys in exactly the order written: $\boxed{7}$ followed by $\boxed{\div}$ followed by $\boxed{4}$. Keying sequences for numbers are abbreviated. For instance, 9.31 is represented by $\boxed{9.31}$ rather than $\boxed{9}$ $\boxed{\bullet}$ $\boxed{3}$ $\boxed{1}$.

## 2.3  How a Graphing Calculator Draws the Graph of a Function

A graphing calculator draws the graph of a function in much the same manner that paper-and-pencil graphs are produced. It plots points of the form $(x, f(x))$. To plot a graph, a graphing calculator needs a function $y = f(x)$, a minimum input value (Xmin), and a maximum input value (Xmax). It uses the Xmin and Xmax values to obtain a large but finite set of input values ($x$) to substitute into the given function $f$ to determine the output values ($f(x)$).

A graphing calculator typically tries to plot as many points as there are columns of pixels in its graphics window. Assume your graphing calculator has 96 columns of pixels (as is the case with the TI-81). We will denote the 96 $x$-values that the graphing calculator tries to substitute into $f$ as $x_0, x_1, x_2, x_3, \ldots, x_i, \ldots, x_{95}$. The first $x$-value, $x_0$, always equals Xmin, and the last $x$-value, $x_{95}$, always equals Xmax. To determine the other 94 $x$-values, the graphing calculator first computes the difference between consecutive $x$'s, that is, the change in $x$ required to go from one $x$-value to the next. We denote this change in $x$ by $\Delta x$, read "delta $x$" or "the change in $x$." The change in $x$ is computed using the following formula:

$$\Delta x = \frac{\text{Xmax} - \text{Xmin}}{95} \qquad (1)$$

The set of $x$-values to be substituted into the function $f$ is the following:

$$x_0 = \text{Xmin}, \quad x_1 = \text{Xmin} + \Delta x, \quad x_2 = \text{Xmin} + 2\Delta x, \quad x_3 = \text{Xmin} + 3\Delta x, \ldots,$$
$$x_i = \text{Xmin} + i\Delta x, \ldots, \quad x_{95} = \text{Xmin} + 95\Delta x = \text{Xmax}.$$

A graphing calculator computes a functional value $y_i = f(x_i)$ for each of these $x_i$ beginning with $x_0$, and immediately plots the point $(x_i, y_i)$ in the graphics window and may (depending on the calculator and mode) connect it to the preceding point, if any, by a line segment. If any one of these $x$-values causes the function $f$ to be undefined *or yields a y-value that is off the screen*, the graphing calculator moves on to the next $x$-value and repeats the process.

## 2.4  The Discrete Nature of Graphing Calculators

Graphing calculators use a *finite* number of points to represent the graph of a function that usually consists of an *infinite* collection of points. When we look at a calculator generated graph, we see a representation that is only suggestive of the actual graph. It is impossible to plot *all* of the points belonging to the graph of most functions. Sometimes important behavior of a function is "hidden" from view unless the graph is magnified many times. There are many examples of hidden behavior in the Demana, Waits, and Clemens textbooks.

## 2.5 Error

Determining the size of $\Delta x$ defined by (1) in Section 2.3 provides a mathematically sound and accurate **estimate** of the error in using a calculator-displayed $x$-value as a graphic solution to an equation. To determine the size of $\Delta x$, you can use a formula like (1), or simply pay close attention to which digits change in the $x$-coordinate readout, and by how much, as you *Trace* from point to point.

If you need to estimate the error in both the $x$- *and* $y$-coordinates of a point on the screen as a solution, then you need to know the number of $y$-values used by the calculator. Assume your graphing calculator has 64 rows of pixels (as in the case with the TI-81). The 64 $y$-values can be denoted by $y_0$, $y_1$, ..., $y_{63}$, where $y_0 = Y\min$ and $y_{63} = Y\max$. To determine the other 62 $y$-values, the graphing calculator first computes the difference between consecutive $y$'s, that is, the change in $y$ required to go from one $y$-value to the next. We denote this change by $\Delta y$, read "delta $y$" or "the change in $y$," and its value is given by the formula:

$$\Delta y = \frac{Y\max - Y\min}{63} \qquad (2)$$

Thus, we have the following representation for the $y$-values:

$y_0 = Y\min, \ y_1 = Y\min + \Delta y, \ y_2 = Y\min + 2\Delta y, \ldots, y_j = Y\min + j\Delta y, \ldots,$
$y_{63} = Y\min + 63\Delta y = Y\max.$

An accurate **estimate** of the error in using a given $y$-coordinate in a solution is $\Delta y$. The set of coordinates $(x_i, y_j)$, where $x_i = X\min + i\Delta x$ and $y_j = Y\min + j\Delta y$, are the "screen coordinates" of points on the display of a graphing calculator.

If you use the *Trace* feature, the $y$-values displayed are actual functional values, not the "screen $y$-coordinates" $y_j$ as defined in the previous paragraph. Their successive differences give approximations to the error in $y$, the function values. In actual examples, you will see that these successive differences can vary quite a bit as you trace across the screen. The differences also depend on the value of $x$ as well as the error in the $x$-value.

## 2.6 Order of Operations

Most graphing calculators use a priority sequence similar to a standard algebraic hierarchy called AOS to determine the order of operations when performing computations. So, for example, it performs the operations of multiplication and division before those of addition and subtraction. Read your calculator *Owner's Manual* for the full details concerning order of operations on your specific calculator.

## 2.7 Using the "Scientific" Functions

Like any scientific calculator, graphing calculators have many useful built-in functions. But the order in which you press the keys on some graphing calculators differs from most traditional scientific calculators. Here are some examples to get you started using the scientific functions. Once you gain control of these functions, try exploring other keys on your own.

**Powers and Roots.** Clear your calculator's screen. Perform the following three computational examples. Figure 2.1 shows how these computations should appear on typical AOS graphing calculator screens. Refer to the figure as you proceed through the computations.

(1)  Evaluate $\sqrt{81}$ by pressing $\boxed{\sqrt{\ }}$ $\boxed{81}$ $\boxed{\text{ENT}}$.

(2)  Compute $5^2$ by keying in $\boxed{5}$ $\boxed{x^2}$ $\boxed{\text{ENT}}$.

(3)  Evaluate $(-5)^4$ using $\boxed{(}$ $\boxed{(-)}$ $\boxed{5}$ $\boxed{)}$ $\boxed{\wedge}$ $\boxed{4}$ $\boxed{\text{ENT}}$. Remember not to confuse the additive inverse, or "sign change," key $\boxed{(-)}$ with the subtraction operation key $\boxed{-}$. Also the exponentiation symbol $\boxed{\wedge}$ is $\boxed{x^y}$ on some calculators.

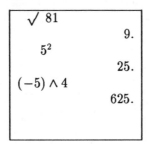

**Figure 2.1.** Powers and roots on the TI-81.

# Chapter 3

---

## The TI-81 Graphing Calculator

The TI-81 is a second generation graphing calculator. It is unique in its user friendly yet powerful functionality. The keys of the TI-81 are grouped by color and physical layout to allow easy location of the key you need. The keys are divided into four zones: graphing keys, editing keys, advanced function keys, and scientific calculator keys (see Figure 3.1). We will describe the keys by rows.

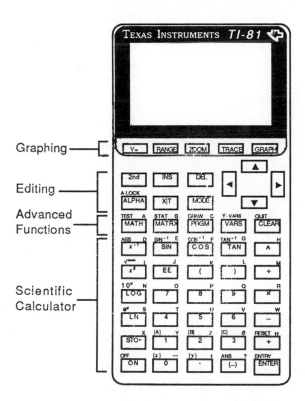

**Figure 3.1**

(1) The **Graphing Keys** (first row) are most frequently used to access the interactive graphing features of the TI-81.

(2) The **Editing Keys** (second and third row) are mostly used for editing expressions and values.

(3) The **Advanced Function Keys** (fourth row) are used to access the advanced functions of the TI-81 through full-screen menus.

(4) The **Scientific Calculator Keys** (the fifth and remaining rows) are used to access the capabilities of a standard scientific calculator.

## 3.1 Getting Started on the TI-81

**3.1.1 Setting the Display Contrast:** Turn the TI-81 on by pressing $\boxed{\text{ON}}$. The brightness and contrast of the display depend on room lighting, battery freshness, viewing angle, and adjustment of the display contrast. The contrast setting is retained in memory when the TI-81 is turned off. You can adjust the display contrast to suit your viewing angle and lighting conditions at any time. As you change the contrast setting, the display contrast changes, and a number in the upper right corner between 0(lightest) and 9(darkest) indicates the current contrast setting.

To adjust the contrast:

(1) Press and release the $\boxed{\text{2nd}}$ key.

(2) Use one of two keys:

- To increase the contrast to the setting that you want, press *and hold* $\boxed{\triangle}$.

- To decrease the contrast to the setting that you want, press *and hold* $\boxed{\triangledown}$.

*Caution* If you adjust the contrast setting to zero, the display may become completely blank. If this happens, press $\boxed{\text{2nd}}$ and then press and hold $\boxed{\triangle}$ until the display reappears. When the batteries are low, the display begins to dim (especially during calculations), and you must adjust the contrast to a higher setting. If you find it necessary to set the contrast to a setting of 8 or 9, you should replace the batteries soon.

**3.1.2 The Display of the TI-81:** The TI-81 displays both text and graphs. When text is displayed, the screen can display up to eight lines of 16 characters per line. When all eight lines of the screen are filled, text "scrolls" off the top of the screen. When you turn the TI-81 on, the *Home screen* is displayed. The Home screen is the primary screen of the TI-81. On it you enter expressions and instructions and see the results. The TI-81 has several types of cursors. In most cases, the appearance of the cursor indicates what will happen when you press the next key. The cursors that you see on the Home screen are listed in Table 3.1. Other special cursors are described later.

When the TI-81 is calculating or graphing, a small box ("Busy" indicator) in the upper right of the screen is highlighted. You can return to the Home screen from any other screen by pressing $\boxed{\text{2nd}}$ $\boxed{\text{QUIT}}$.

| Cursor | Appearance | Meaning |
|--------|-----------|---------|
| Entry cursor | Solid blinking rectangle | The next keystroke is entered at the cursor overwriting any character |
| Insert cursor | Blinking underline cursor | The next keystroke is inserted at the cursor |
| 2nd cursor | Highlighted blinking ↑ | The next keystroke is a second function |
| ALPHA cursor | Highlighted blinking **A** | The next keystroke is an alpha character |

**Table 3.1**

**3.1.3  Entering a Calculation:** To begin, enter $1000(1.06)^{10}$ , by keying ⌊1000⌋ ⌊×⌋ ⌊1.06⌋ ⌊∧⌋ ⌊10⌋ just as you would write it down.

The entire expression is shown in the first line of the display. Press ⌊ENTER⌋ to evaluate the expression. The result of the expression is shown on the right side of the second line on the display (see Figure 3.2). The cursor is positioned on the left side of the third line, ready for you to enter the next expression.

```
1000 * 1.06 ∧ 10
            1790.847697


```

**Figure 3.2**

Notice the difference between this display and a typical scientific calculator. You can see the complete problem *and* the solution!

**3.1.4  The Menu (Advanced Functions) Keys:** You can access functions and operations that are not on the keyboard through *menus*. A menu screen temporarily replaces the screen where you are working. After you select an item from a menu, the screen where you are working is displayed again. The MATH menu is shown in Figure 3.3. The down arrow ↓ at the bottom of the ⌊MATH⌋ column indicates at least one additional option is available that does not appear on the screen. It can be viewed by pressing the ⌊▽⌋ key seven times.

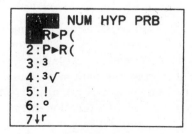

**Figure 3.3**

The MATH menu screen contains four columns of menus. Press ▷ slowly to see all four columns of menus. A particular item is selected by typing the number of the item desired. For example, to compute 7! key ⑦ MATH [5: !] ENTER (see Figure 3.4).

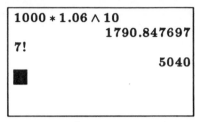

**Figure 3.4**

**Numbered-menu Notation.** For numbered menus, like the MATH menu, we adopt a special notation that is a modification of the keying sequence notation introduced in Section 2.1. For instance, if your calculator screen looks like Figure 3.3 and you wish to select the factorial function (!) from the MATH menu, we will denote the keystroke using [5:!].

**3.1.5 The Mode Screen:** Press MODE to change the modes of the TI-81. The default modes are the first column of choices. To make a selection, use the arrow keys to move the cursor to the desired row first and then the desired column. Then press ENTER to select the item. For example, to select "Dot" mode, press the down arrow ▽ key four times and ▷ once. Then press ENTER to select "Dot" mode.

The first row contains the Normal, Scientific, or Engineering notation display settings. Notation formats affect only how a numeric result is displayed. You can enter a number in any format. Normal display format is the way in which we usually express numbers, with digits to the left and right of the decimal point, as in 12345.67. Scientific notation expresses numbers in two parts. The significant digits are displayed with one digit to the left of the decimal point. The appropriate power of 10 is displayed to the right of $E$, as in 1.234567E4. Engineering notation is similar to scientific notation. However, the number may have one, two, or three digits before the decimal point, and the power-of-10 exponent is a multiple of three, as in 12.34567E3.

The second row contains the floating or fixed decimal display settings.

The third row contains the radian or degree angle settings. Radian setting means that angle arguments in trig functions or polar-rectangular conversions are interpreted as radians. Results are displayed in radians. Degree setting means that angle argument in trig functions or polar-rectangular conversions are interpreted as degrees. Results are displayed in degrees.

The fourth row contains the function or parametric graphing settings. Function graphing plots a function where $Y$ is expressed in terms of $X$. See Section 3.3 for more information about function graphing. Parametric graphing plots a relation where $X$ and $Y$ are each expressed in terms of a third variable, T. See Section 3.5 for more information about graphing parametric equations.

The fifth row contains the connected line or dot graph display settings. A connected line graph draws a line between the points calculated on the graph of a function in the $Y =$ list. A dot graph plots only the calculated points on the graph.

The sixth row contains the sequential or simultaneous plotting settings. Sequential plotting means that, if more than one function is selected, one function is evaluated and plotted completely before the next function is evaluated and plotted. Simultaneous plotting means that, if more than one function is selected, all functions are evaluated and plotted for a single value of $X$ or $T$ before the functions are evaluated and plotted for the next value of $X$ or $T$.

The seventh row are the grid off or grid on setting. Grid Off means that no grid points are displayed on a graph. Grid On means that grid points are displayed on a graph. Grid points correspond to the axes tick marks.

The last row contains the rectangular or polar coordinate display settings. Rectangular coordinate display shows the cursor coordinate at the bottom of the screen in terms of rectangular coordinates $X$ and $Y$. Polar coordinate display shows the cursor coordinate at the bottom of the screen in terms of polar coordinates $R$ and $\theta$.

**3.1.6 Leaving a Menu or Edit Screen:** There are several ways to leave a menu or edit screen. After you make a selection from a menu, you usually are returned to the screen where you were. If you decide not to make a selection from a menu, you can leave the menu in one of the following ways:

(1)  Press $\boxed{\texttt{2nd}}$ $\boxed{\texttt{QUIT}}$ to return to the Home screen.

(2)  Press $\boxed{\texttt{CLEAR}}$ to return to the screen where you were.

(3)  Select another screen by pressing the appropriate key, such as $\boxed{\texttt{MATH}}$ or $\boxed{\texttt{RANGE}}$.

When you finish entry or editing tasks, such as entering range values, entering statistical data, editing a program, or changing modes, leave the menu in one of the following ways:

(1)  Press $\boxed{\texttt{2nd}}$ $\boxed{\texttt{QUIT}}$ to return to the Home screen.

(2)  Press another edit screen key, such as $\boxed{\texttt{RANGE}}$.

## 3.2  Calculations on the TI-81

**3.2.1 Last Entry:** When $\boxed{\texttt{ENTER}}$ is pressed on the Home screen and an expression is evaluated successfully, the TI-81 stores the current expression in a special storage area called Last Entry. It can be recalled by pressing $\boxed{\texttt{2nd}}$ $\boxed{\texttt{ENTRY}}$ or the up arrow key $\boxed{\triangle}$.

Because the TI-81 updates the Last Entry storage area only when $\boxed{\texttt{ENTER}}$ is pressed, you can recall the last entry even if you have begun entering the next expression. However, the Last Entry overwrites the current expression. When you turn the TI-81 off, the expression in Last Entry is retained in memory.

*Example*   Determine when an investment earning interest at 8.5% compounded monthly will double in value. The applicable compound interest equation is $2 = \left(1 + \frac{.085}{12}\right)^n$. We want to solve this equation for $n$. Make a guess, say $n = 100$. Key in $\boxed{(}\,\boxed{1}\,\boxed{+}\,\boxed{.085}\,\boxed{\div}\,\boxed{12}\,\boxed{)}\,\boxed{\wedge}\,\boxed{100}$.

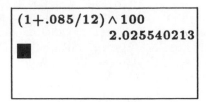

**Figure 3.5**

Make a new estimate based on the above result (see Figure 3.5). The new estimate should be smaller than 100. Why? Make your next estimate $n = 99$ and use the last entry feature. Press the $\boxed{\triangle}$ key.
Notice the cursor is at the end of the expression (see Figure 3.6). Move the cursor 3 spaces left and type $\boxed{99}$. Press the $\boxed{\text{DEL}}$ key to delete the "0". Now press $\boxed{\text{ENTER}}$ to obtain the next result (see Figure 3.7).

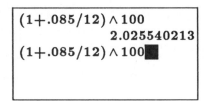

**Figure 3.6**

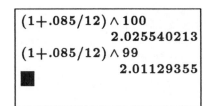

**Figure 3.7**

Continue estimating in this manner until your answer is as accurate as you wish.

**3.2.2 Last Answer:** The $\boxed{\text{2nd}}$ $\boxed{\text{ANS}}$ key recalls the last answer in a computation. For example, if you just computed the example in 3.2.1 do the following. Turn your TI-81 off. Turn it on. Press $\boxed{\text{2nd}}$ $\boxed{\text{ANS}}$ $\boxed{\text{ENTER}}$. Notice the last answer is stored as a variable (Ans) that can be used in computations.

## 3.3  Graphing on the TI-81

The keys on the TI-81 that are related most closely to graphing are located immediatly under the display. When you press $\boxed{\text{Y=}}$, an edit screen is displayed where you enter and select the functions that you want to graph. When you press $\boxed{\text{RANGE}}$, an edit screen is displayed where you define the viewing rectangle for the graph. When you press $\boxed{\text{ZOOM}}$, you access a menu of instructions that allow you to change the viewing rectangle. When you press $\boxed{\text{TRACE}}$, you can move the cursor along a graphed function and display the $X$ and $Y$ coordinate values of the cursor location on the function. When you press $\boxed{\text{GRAPH}}$, a graph of the currently selected functions is displayed in the chosen viewing rectangle.

**3.3.1  Entering a Function:** Press $\boxed{\text{Y=}}$. The display shows labels for four functions. The cursor is positioned at the beginning of the first function. Enter the two function $f(x) = x^3 - 2x$ and $g(x) = 2\cos x$ so that $y_1 = f(x)$ and $y_2 = g(x)$. To enter the first expression press $\boxed{\text{CLEAR}}$ $\boxed{\text{X/T}}$ $\boxed{\wedge}$ $\boxed{3}$ $\boxed{-}$ $\boxed{2}$ $\boxed{\text{X/T}}$, then press $\boxed{\text{ENTER}}$ to move the cursor to the next function. (*Note:* The $\boxed{\text{X/T}}$ key lets you enter the variable $X$ quickly without pressing $\boxed{\text{ALPHA}}$.) Notice the = sign is highlighted and thickened to show that $Y_1$ is "selected" to be graphed (see Figure 3.8). Now enter the second expression by pressing $\boxed{\text{CLEAR}}$ $\boxed{2}$ $\boxed{\text{COS}}$ $\boxed{\text{X/T}}$ $\boxed{\text{ENTER}}$. Pressing $\boxed{\text{ENTER}}$ with the cursor on the equals sign will change the status (selected or unselected) of the function.

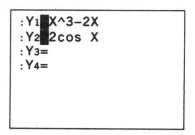

**Figure 3.8**

**3.3.2 Checking the Viewing Rectangle:** The Range key allows you to choose the viewing rectangle that defines the portion of the coordinate plane that appears in the display. The values of the RANGE variables determine the size of the viewing rectangle and the scale units for each axis. You can view and change the values of the RANGE variables almost any time. Press RANGE to display the RANGE variables edit screen (see Figure 3.9). The values shown here on the RANGE edit screen are the standard default values $[-10, 10]$ by $[-10, 10]$. Xscl and Yscl give the distance between consecutive tick marks on the coordinate axes. Both are 1 in Figure 3.9. To change one of the RANGE values, move the cursor to the desired line and type in the new value. You may also edit an existing value to produce a new value. Xres controls the number of pixels used to obtain a plot. We will normally use Xres=1, which gives the best possible resolution.

```
RANGE
Xmin=-10
Xmax=10
Xscl=1
Ymin=-10
Ymax=10
Yscl=1
Xres=1
```

**Figure 3.9**

**Displaying the Graph.** Press GRAPH to graph the selected functions ($f$ and $g$) in the current viewing rectangle with default mode settings (see Figure 3.10). When the plotting is completed, press ▷ once to display the graphics cursor just to the right of the center of the screen (see Figure 3.11). The bottom line in the display shows *both* the $X$ and $Y$ coordinate values for the position of the graphics cursor. Use the cursor-movement keys (◁, ▷, △, ▽), to move the cursor. As you move the cursor, the $X$ and $Y$ coordinate values are updated continually with the cursor position.

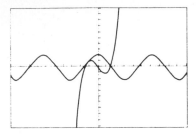

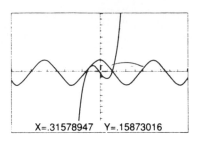

X=.31578947    Y=.15873016

**Figure 3.10**                                        **Figure 3.11**

## 3.4  Zooming In on the Graph

You can magnify a portion of the viewing rectangle around a specific location by selecting the *Zoom In* instruction from the ZOOM menu. Press ZOOM to access the menu of built-in ZOOM functions (see Figure 3.12).

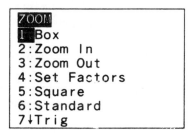

**Figure 3.12**

This menu is typical of all numbered menus of built-in operations on the TI-81. To select an item from a menu, either press the number to the left of the instruction you want, or press ▽ to position the cursor on that instruction and then press ENTER. *Note:* It is much easier and quicker simply to press the *number* of the desired action. To zoom in, press 2 to select the Zoom In instruction from the menu. The graph is displayed again. Notice the cursor has changed to indicate that you are using a ZOOM instruction. Place the cursor at the point indicated in Figure 3.13, press ENTER. The current position of the cursor becomes the center of the new viewing rectangle. The new viewing rectangle has been adjusted in both the *X* direction and the *Y* direction by factors of 4, which are the default values for the zoom factors (see Figure 3.14). The zoom factors can be changed by pressing ZOOM [4: Set Factors] (see Section 3.4.5).

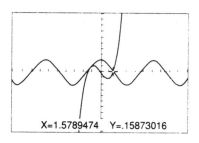

**Figure 3.13**

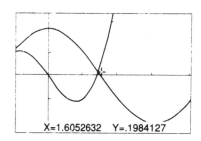

**Figure 3.14**

**3.4.1  Moving between the Graph and Range Screens:**  When the TI-81 executes a ZOOM instruction, it updates the RANGE variables to reflect the new viewing rectangle.  You can check the RANGE values by pressing $\boxed{\text{RANGE}}$ to see the size of the new viewing rectangle and then return to the graph by pressing $\boxed{\text{GRAPH}}$ without having to replot the graph.  The modified values resulting from the Zoom In instruction depend on the exact cursor position when you executed the Zoom In instruction.  Press $\boxed{\text{GRAPH}}$ to see the graph again at any time.

**3.4.2  Moving the Cursor along a Function:**  The TRACE feature allows you to move the cursor along a graph showing the $x$ and $f(x)$ coordinate values of the cursor location on the graphs.  Press $\boxed{\text{TRACE}}$. The cursor appears near the middle of the screen on the $Y_1 = X^3 - 2X$ function (see Figure 3.15).  The keys $\boxed{\triangleright}$ and $\boxed{\triangleleft}$ allow you to move along the graph.  The coordinate values of the cursor location are displayed at the bottom of the screen.  The $Y$ value shown is the calculated function value $f(x)$. If the cursor moves off the top or bottom of the screen, the coordinate values $X$ and $Y$ displayed at the bottom of the screen continue to change appropriately.  Panning is possible in function graphing.  Moving the cursor by pressing the right or left arrow keys a sufficient number of times will cause the graph to pan to the right or left, respectively.

Press $\boxed{\triangledown}$. The cursor moves to the next active function, in this case, $Y_2 = 2\cos x$, at the same $X$ value where it was located on the first function (see Figure 3.16).  $\boxed{\triangle}$ and $\boxed{\triangledown}$ allow you to move among all functions that are defined *and* selected (active).

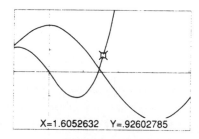

**Figure 3.15**

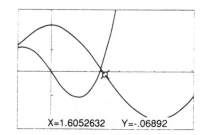

**Figure 3.16**

**3.4.3  Using Zoom Box:**  There is another way to magnify a graph. The Zoom [1: Box] Instruction lets you adjust the viewing rectangle by drawing a box on the display to define the new viewing rectangle.  To adjust the viewing rectangle to the standard default range, press $\boxed{\texttt{ZOOM}}$ [6: Standard]. This automatically adjusts the viewing rectangle to the standard default viewing rectangle $[-10, 10]$ by $[-10, 10]$. This display shows the same graph that you saw earlier (see Figure 3.13). Press $\boxed{\texttt{ZOOM}}$ [1: Box]. This lets you draw a box anywhere on the screen in which to magnify the graph for a new viewing. The cursor is in the middle of the screen. Its new appearance indicates that you have selected a ZOOM instruction. Use the cursor arrow keys to move the cursor from the middle of the graph to where you *want one corner* of the new viewing rectangle to be (see Figure 3.17). Press $\boxed{\texttt{ENTER}}$.

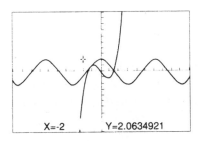

**Figure 3.17**

Notice that the cursor has changed to a small box. Next, move the cursor to the *diagonally opposite corner* of the desired viewing rectangle (see Figure 3.18). The outline of the new viewing rectangle is drawn as you move the cursor. Press $\boxed{\texttt{ENTER}}$ to accept the cursor location as the second corner of the box. The graph is replotted immediately using the box outline as the new viewing rectangle. Use a second zoom-in to obtain a picture like the one shown in Figure 3.19.

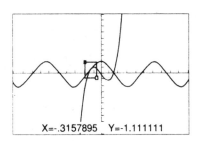

**Figure 3.18**

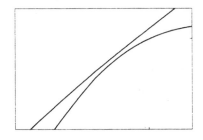

**Figure 3.19**

To leave the graph display, press $\boxed{\texttt{2nd}}$ $\boxed{\texttt{QUIT}}$ or $\boxed{\texttt{CLEAR}}$ to return to the Home screen.

**3.4.4 Exploring a Graph with the Free-Moving Cursor:** After pressing $\boxed{\texttt{GRAPH}}$, the free-moving cursor can be moved to identify the coordinate of any location on the graph. You can use $\boxed{\triangleleft}$, $\boxed{\triangleright}$, $\boxed{\triangle}$, and $\boxed{\triangledown}$ to move the cursor around the graph. When you first press $\boxed{\texttt{GRAPH}}$ to display the graph, no cursor is visible. As soon as you press one of the cursor-movement keys, the cursor moves from the center of the viewing rectangle. As you move the cursor around the graph, the coordinate values of the cursor location are displayed at the bottom of the screen. Coordinate values always appear in floating-decimal format. The numeric display settings on the MODE screen do not affect coordinate display.

In Rectangular mode, moving the cursor updates and displays the values of the rectangular coordinates $X$ and $Y$. In Polar mode, the coordinates $R$ and $\theta$ are updated and displayed. To see the graph without the cursor or coordinate values, press $\boxed{\texttt{GRAPH}}$ or $\boxed{\texttt{ENTER}}$. When you press a cursor-movement key, the cursor moves from the middle of the viewing rectangle again if you pressed $\boxed{\texttt{GRAPH}}$ or from the same point if you pressed $\boxed{\texttt{ENTER}}$.

*Note*   The free-moving cursor moves from dot to dot on the screen. When you move the cursor to a dot that appears to be "on" the function, it may be near, but not on, the function; therefore, the coordinate value displayed at the bottom of the screen is not necessarily a point on the function. The coordinate value is accurate to within the width of the dot (see Section 2.5). To move the cursor along a function, use the TRACE feature.

**3.4.5  Setting Zoom Factors:** Zoom factors determine the scale of the magnification for the Zoom In or Zoom Out features. Before using Zoom In or Zoom Out, you can review or change the zoom factors. Zoom factors are positive numbers (not necessarily integers) greater than or equal to 1.

To review the current values of the zoom factors, select [4: Set Factors] from the ZOOM menu. The ZOOM FACTORS screen appears. Figure 3.20 shows how the ZOOM FACTORS editing menu appears with the default factors of 4 in each direction.

```
ZOOM FACTORS
XFact=4
YFact=4
```

**Figure 3.20**

If the factors are not what you want, change them in one of the following ways:

(1)  Enter a new value. The original value is cleared automatically when you begin typing.

(2)  Position the cursor over the digit you want to change. Then type over it or use $\boxed{\texttt{DEL}}$ to delete it.

When the zoom factor values are as you want them, leave ZOOM FACTORS in one of the following ways:

(1)  Select another screen by pressing the appropriate key, such as $\boxed{\texttt{GRAPH}}$ or $\boxed{\texttt{ZOOM}}$.

(2)  Press $\boxed{\texttt{2nd}}$ $\boxed{\texttt{QUIT}}$ to return to the Home screen.

**3.4.6  Using Zoom Out:** Zoom Out displays a greater portion of the graph, centered around the cursor location, to provide a more global view. The XFact and YFact settings determine the extent of the zoom. After checking or changing the zoom factors select [3: Zoom Out] from the ZOOM menu. Notice the special cursor. It indicates that you are using a Zoom instruction.

Move the cursor to the point that you want as the center of the new viewing rectangle and then press $\boxed{\texttt{ENTER}}$. The TI-81 adjusts the viewing rectangle by **XFact** and **YFact**, updates the RANGE variables, and replots the selected functions, centered around the cursor lcoation.

To zoom out again:

(1)  Centered at the same point, press $\boxed{\text{ENTER}}$ .

(2)  Centered at a new point, move the cursor to the point that you want as the center of the new viewing
rectangle and then press $\boxed{\text{ENTER}}$ .

When you finish using the Zoom Out feature, leave in one of the following ways:

(1)  Select another screen by pressing the appropriate key, such as $\boxed{\text{TRACE}}$ or $\boxed{\text{GRAPH}}$ .

(2)  Press $\boxed{\text{2nd}}$ $\boxed{\text{QUIT}}$ or $\boxed{\text{CLEAR}}$ to return to the Home screen.

**3.4.7  Using Other ZOOM Features:**  Four of the ZOOM features reset the RANGE variables to
predefined values or use factors to adjust the RANGE variables. Xres remains unchanged, except in Standard.

(1)  **Square.**  The TI-81 replots the functions, redefining the viewing rectangle using values based on the
current RANGE variables, but adjusted to equalize the width of the dots on the X and Y axes. **Xscl**
and **Yscl** remain unchanged. This feature makes the graph of a circle look like a circle (see Section
3.4.8).

The TI-81 replots the graph as soon as the menu selection is made. The center of the current graph
becomes the center of the new graph.

(2)  **Standard.**  The TI-81 updates the RANGE variables to the standard default values and replots the
graph as soon as the menu selection is made. The RANGE variable standard defaults are:

$$\text{Xmin} = -10 \qquad \text{Ymin} = -10 \qquad \text{Xres} = 1$$
$$\text{Xmax} = 10 \qquad \text{Ymax} = 10$$
$$\text{Xscl} = 1 \qquad \text{Yscl} = 1$$

(3)  **Trig.**  The TI-81 updates the RANGE variables using preset values appropriate for trig functions and
replots the graph as soon as the menu selection is made. The trig RANGE variable values in **Radians**
mode are:

$$\text{Xmin} = -2\pi \qquad \text{Ymin} = -3$$
$$\text{Xmax} = 2\pi \qquad \text{Ymax} = 3$$
$$\text{Xscl} = \pi/2 \qquad \text{Yscl} = .1$$

*Note*  The display shows the numeric values of $2\pi$ , 6.283185307 , and $\pi/2$ , 1.570796327 .

(4)  **Integer.**  When you select [8: Integer] from the ZOOM menu, you can move the cursor to the point
that you want as the center of the new viewing rectangle and then press $\boxed{\text{ENTER}}$ .

The TI-81 replots the functions, redefining the viewing rectangle so that the mid-point of each dot on
the X and Y axis is an integer. **Xscl** and **Yscl** are equal to 10.

**3.4.8  Example: Graphing a Circle**

*Problem*  Graph a circle of radius 10, centered around the origin $x^2 + y^2 = 10$ .

*Solution*  To graph a circle, you must enter separate formulas for the upper and lower portions of the circle.
Use Connected Line mode.

(1) Press $\boxed{\texttt{Y=}}$. Enter the expressions to define two functions. The top half of the circle is defined by $Y_1 = \sqrt{100 - X^2}$ The bottom half of the circle is defined by $Y_2 = -Y_1$ (key $\boxed{\texttt{( - )}}$ $\boxed{\texttt{2nd}}$ $\boxed{\texttt{Y-Vars}}$ [1: $Y_1$] (use the Y-VARS menu)).

(2) Press $\boxed{\texttt{ZOOM}}$ [6: Standard]. This is a quick way to reset the RANGE variables to the standard defaults. It also graphs the functions, so you do not need to press $\boxed{\texttt{GRAPH}}$. Notice that the graph appears to be an ellipse (see Figure 3.21).

(3) To adjust the display so that each "dot" has an equal width and height, press $\boxed{\texttt{ZOOM}}$ and then select [5: Square]. The functions are replotted and now appear as a circle on the display (see Figure 3.22).

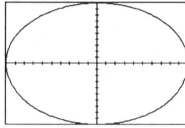
**Figure 3.21**

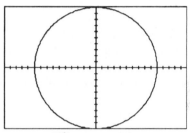
**Figure 3.22**

(4) To see the effect of the Zoom Square instruction on the RANGE variables, press $\boxed{\texttt{RANGE}}$ and notice the values for Xmin, Xmax, Ymin, and Ymax.

## 3.5  Parametric Graphing

**3.5.1  Defining and Displaying a Parametric Graph:**  Parametric equations consist of an $X$ component and a $Y$ component, each expressed in terms of the same independent variable $T$. Up to three pairs of parametric equations can be defined and graphed at a time. The steps for defining a parametric graph are the same as those for defining a function graph. Differences are noted below.

Press $\boxed{\texttt{MODE}}$ to display the MODE settings (see Figure 3.23). The current settings are highlighted. To graph parametric equations, you must select 〈Param〉 (see Section 3.1.5) before you enter RANGE variables or enter the components of parametric equations. Also, you usually should select 〈Connected〉 to obtain a more meaningful parametric graph.

Press $\boxed{\texttt{Y=}}$ to display the Y= edit screen (see Figure 3.24).

| Setting | Meaning |
|---|---|
| **Norm** Sci Eng | Type of notation for display |
| **Float** 0123456789 | Number of decimal places |
| **Rad** Deg | Type of angle measure |
| **Function** Param | Function or parametric graphing |
| **Connected** Dot | Whether to connect plotted points |
| **Sequence** Simul | How to plot selected functions |
| **Grid Off** Grid On | Whether to display a graph grid |
| **Rect** Polar | Type of graph coordinate display |

**Figure 3.23**

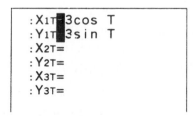

**Figure 3.24**

On this screen, you display and enter both $X$ and $Y$ components. There are three pairs of components, each defined in terms of $T$.

To enter the two expressions that define new parametric equations, follow the procedure in Section 3.3.1.

(1) You must define both the $X$ and $Y$ components in a pair.

(2) The independent variable in each component must be $T$. You may press the $\boxed{\text{X/T}}$ key, rather than pressing $\boxed{\text{ALPHA}}$ $\boxed{\text{T}}$, to enter the parametric variable $T$. (Parametric mode defines the independent variable as $T$.)

The procedures for editing, clearing, and leaving are the same as for function graphing.

Only the parametric equations you select are graphed. You may select up to three equations at a time. Press $\boxed{\text{Y=}}$ to display the $\mathbf{Y=}$ edit screen to select and unselect equations. The = signs on the selected pairs of equations are highlighted.

To change the selection status of a parametric equation:

(1) Place the cursor over the = sign on either the $X$ or $Y$ component.

(2) Press $\boxed{\text{ENTER}}$ to change the status. The status on both the $X$ and $Y$ components is changed.

*Note* When you enter or edit either component of an equation, that pairs of equations is selected automatically.

Press $\boxed{\text{RANGE}}$ to display the current RANGE variable values. The values shown in Table 3.2 are the *standard defaults in Radian mode*. Notice that **Xres**, which appeared on the function graphing RANGE edit screen, is not here; but three new variables, **Tmin**, **Tmax**, and **Tstep**, are.

| Setting | Meaning |
|---|---|
| RANGE | |
| Tmin=0 | The smallest $T$ value to be evaluated |
| Tmax=2 $\pi$ | The largest $T$ value to be evaluated |
| Tstep = $\pi/30$ | The increment between $T$ values |
| Xmin= $-10$ | The smallest $X$ value to be displayed |
| Xmax=10 | The largest $X$ value to be displayed |
| Xscl=1 | The spacing between $X$ tick marks |
| Ymin= $-10$ | The smallest $Y$ value to be displayed |
| Ymax=10 | The largest $Y$ value to be displayed |
| Yscl=1 | The spacing between $Y$ tick marks |

**Table 3.2**

*Note* The display shows the numeric value of $2\pi$, 6.283185307, for **Tmax** and 0.104719755 for **Tstep**.

To change the value of a RANGE variable or to leave the screen, follow the procedures in Section 3.3.2.

When you press $\boxed{\text{GRAPH}}$, the TI-81 plots the selected parametric equations. It evaluates both the X and the Y component for each value of T (from **Tmin** to **Tmax** in intervals of **Tstep**) and then plots each point defined by X and Y. The RANGE variables define the viewing rectangle. As a graph is plotted, the TI-81 updates the coordinates X and Y and the values of the parameter T.

**3.5.2  Exploring a Parametric Graph:** As in function graphing, three tools are available for exploring a graph: using the free-moving cursor, tracing an equation, and zooming. The free-moving cursor works in parametric graphing just as it does in function graphing. The cursor coordinate values for X and Y (or R and $\theta$ in polar mode) are updated and displayed.

The TRACE feature (see Section 3.4.2) lets you move the cursor along the equation one **Tstep** at a time. When you begin a trace, the blinking cursor is on the first selected equation at the middle T value and the coordinate values of X, Y, and T are displayed at the bottom of the screen. As you trace along a parametric graph using $\boxed{\triangleleft}$ $\boxed{\triangleright}$, the values of X, Y and T are updated and displayed. The X and Y values are calculated from T.

If the cursor moves off the top or bottom of the screen, the coordinate values of X, Y, and T displayed at the bottom of the screen continue to change appropriately.

Panning is not possible on parametric curves. To see a section of the equations not displayed on the graph, you must change the RANGE variables.

The ZOOM features work in parametric graphing as they do in function graphing.

Only the X (**Xmin**, **Xmax**, and **Xscl**) and Y (**Ymin**, **Ymax**, and **Yscl**) RANGE variables are affected. The T RANGE variables (**Tmin**, **Tmax**, and **Tstep**) are not affected, except when you select [6: Standard]; in that case they become **Tmin** $= 0$, **Tmax** $= 2\pi$, and **Tstep** $= \pi/30$. You may want to change the T RANGE variable values to ensure that sufficient points are plotted.

**3.5.3  Applications of Parametric Graphing**

**Example 1:  Simulating Motion**

*Problem*   Graph the position of a ball kicked from ground level at an angle of 60° with an initial velocity of 40 ft/sec. (Ignore air resistance.) What is the maximum height, and when is it reached? How far away and when does the ball strike the ground?

*Solution*   If $v_0$ is the initial velocity and $\theta$ is the angle, then the horizontal component of the position of the ball as a function of time is described by

$$X(T) = Tv_0 \cos \theta .$$

The vertical component of the position of the ball as a function of time is described by

$$Y(T) = -167T^2 + Tv_0 \sin \theta .$$

In order to graph the equations,

(1)   Press $\boxed{\text{MODE}}$. Select Parametric, Connected Line, and Degree Mode.

(2) Press $\boxed{\text{Y=}}$. Enter the expressions to define the parametric equation in terms of T.

$X_{1T} = 40\text{T}\cos 60$

$Y_{1T} = 40\text{T}\sin 60 - 16\text{T}^2$

(3) Press $\boxed{\text{RANGE}}$. Set the RANGE variables appropriately for this problem.

| | | |
|---|---|---|
| Tmin=0 | Xmin $= -5$ | Xmin $= -5$ |
| Tmax $= 2.5$ | Xmax $= 50$ | Ymax $= 20$ |
| Tstep $= .02$ | Xscl $= 5$ | Yscl $= 5$ |

(4) Press $\boxed{\text{GRAPH}}$ to graph the position of the ball as a function of time.

(5) Now press $\boxed{\text{TRACE}}$ to explore the graph. When you press $\boxed{\text{TRACE}}$, the values for X, Y, and T are displayed at the bottom of the screen. These values change as you trace along the graph.

Move the cursor along the path of the ball to investigate these values. Notice you have a "stop action" picture at each 0.02 seconds.

### Example 2: Graphing a Polar Equation

*Problem*   Graph the spiral of Archimedes, that is, the curve defined by the polar equation $r = a\theta$.

*Solution*   A polar equation $r = f(\theta)$ can be graphed using the parametric graphing features of the TI-81 by applying the conversion formulas, $X = f(\theta)\cos(\theta)$ and $Y = f(\theta)\sin(\theta)$. Thus, the spiral of Archimedes (with $a = 0.5$) can be expressed parametrically as

$$X = 0.5\theta\cos(\theta)$$

$$Y = 0.5\theta\sin(\theta)$$

Graph the equation using the standard default viewing rectangle, Radian mode, and Connected Line mode.

(1) Press $\boxed{\text{MODE}}$. Select Parametric mode. Choose the defaults for the other modes.

(2) Press $\boxed{\text{Y=}}$. Enter the expressions to define the parametric equation in terms of T.

$X_{1T} = 0.5\text{T}\cos \text{T}$

$Y_{1T} = 0.5\text{T}\sin \text{T}$

(3) Press $\boxed{\text{ZOOM}}$ [6: Standard] to graph the equations in the standard default viewing rectangle.

The graph shows only the first loop of the spiral. This is because the standard default values for the RANGE variables define **Tmax** as $2\pi$.

(4) To explore the behavior of the graph further, press $\boxed{\text{RANGE}}$ and change **Tmax** to 25.

(5)   Press [GRAPH] to display the new graph (see Figure 3.25).

(6)   Try pressing [ZOOM] [5: Square]. What happened? (See Figure 3.26.) Contrast with Figure 3.25.

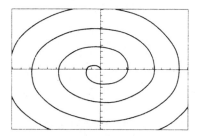

**Figure 3.25**

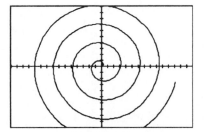

**Figure 3.26**

# Chapter 4

## Introduction to the TI-85 Pocket Computer

### 4.1  The Menu Keys

The TI-85 uses horizontal display menus to give you access to more operations than you can access from the keyboard alone.

#### 4.1.1  The Menus and Menu Keys:

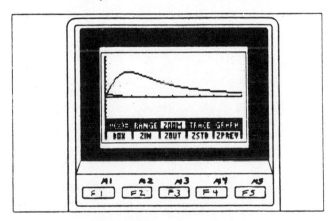

The top of keys are the menu keys and are labeled [F1], [F2], [F3], [F4], and [F5]. The *second* functions of the menu keys are labeled [M1], [M2], [M3], [M4], and [M5]. The menu items are shown on the bottom line(s) of the display, above the five menu keys.

#### 4.1.2  Selecting Menu Items

(1)  To select a menu item from the eighth (bottom) line of the screen display, press the menu key below the item.

(2)  To select a menu item from the seventh (next-to-the-bottom) line of the screen display, press and release [2nd] and then press the menu key below the item.

In this chapter, the **menu** items are indicated by < > brackets. Keys are denoted by [ ] square brackets. For example, press [F2] to select < ZIN > or press [2nd] [M5] to select < GRAPH > .

#### 4.1.3  Finding Commands and Function:  All commands and functions may be found in two or more menus or directly on keys. One menu key you will use often is [2nd] [CATALOG]. It contains <u>all</u> functions

and commands in alphabetical order. For example to select "randM(", simply press the number 5 calculator key becuase that is the gray alpha numeric character "R" (look just above the number 5 key). Notice you can move the pointer down to "randM(" by pressing the down arrow key. Press [ENTER] to make the selection. "RandM)" is also found in a matrix menu.

It is useful to know that all commands, functions, variables, and program names may be *typed directly on the screen* <u>letter by letter</u> using the [ALPHA] key. [2nd] [ALPHA] is lower case. [ALPHA] <u>twice</u> is "caps lock."

**4.1.4   Variables:**  All variables can be found in the [2nd] [VARS] menus. They can be identifed and selected in the same manner as in CATALOG.

**4.1.5   Finding All Menu Items:**  When you press a key and obtain a menu, for example, [GRAPH], notice there is a small ▶ at the end of the visible menu item listing. That means there is MORE! Press the [MORE] key to see 5 more menu listings. Press [MORE] again to see the last 3 menu items in the [GRAPH] menu collection. There are 13 menu items in [GRAPH].

Some [GRAPH] menu items give you access to edit screens like $< y(x)=>$ or $< RANGE >$. Others give you even more menus, like $< ZOOM >$ or $< MATH >$. Finally some menu items are commands like $< GRAPH >$ or $< TRACE >$. The [GRAPH] menu collection is typical of the way many menu keys work like [2nd][MATRIX] or [STAT] or [2nd][MATH].

**4.1.6   Edit Screens:**  <u>Expressions</u> that evaluate to a number may be entered in edit screens. For example in [GRAPH] $< RANGE >$, $2A + \frac{8\pi}{2} - \cos 2.2$ could be directly entered in the xMIN=■ position (assuming A was defined).

## 4.2  Resetting the TI-85

Before beginning these sample problems, follow the steps on this page to ensure that the TI-85 is reset to its factory settings. (Resetting the TI-85 erases all previously entered data.)

(1)   Press [ON] to turn the calculator on.

(2)   Press [2nd] and then press [+]. (Pressing [2nd] accesses the function printed to the left above the next key that you press. MEM is the second function of [+].) The bottom line of the display shows the MEM (memory) menu.

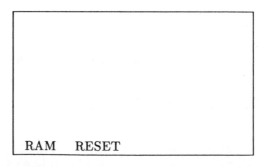

(3)   Press the [F3] menu key to select < RESET >, the third item in the MEM menu. The bottom line is relabeled with the RESET menu and the MEM menu moves up a line.

```
┌─────────────────────────────────────┐
│                                     │
│                                     │
│                                     │
│                                     │
│                                     │
│                                     │
│  RAM    RESET                       │
│  ALL    CLMEM  DFLTS                │
└─────────────────────────────────────┘
```

(4)   Press [F1] to select < ALL >. The display shows the message <u>Are you sure?</u>.

Press [F4] to select < YES >. The display shows the message <u>Mem cleared</u>, and <u>Defaults set</u>.

The display contrast was reset to the default. To adjust the display contrast, press and release [2nd] and then press and hold [ ↑ ] (to make the display darker) or [ ↓ ] (to make the display lighter).

Press [CLEAR] to clear the display.

```
┌─────────────────────────────────────┐
│                                     │
│  Are you sure?                      │
│                                     │
│                                     │
│                                     │
│                                     │
│  CLMEM  DFLTS  ALL                  │
│  NO     YES                         │
└─────────────────────────────────────┘
```

## 4.3  Application: Future Value of an Annuity Due

The TI-85 display can show up to eight lines of 21 characters per line. This lets you see each expression or instruction in its entirety as it is entered. Variable names can be up to eight characters, in uppercase and lowercase. Names are case-sensitive. You can enter more than one command on a line by concatenating them with a : (colon).

If you invest $25 at the beginning of each month at 6% annual interest, compounded monthly, how much money you will have at the end of three years? The formula is:

$$\text{PMT} \frac{(1 + I)^{N+1} - (1 + I)}{I}$$

(1)   To store the value ($25) for the payment amount in the variable PMT, press 25 [STO ▷]. When you press the [STO ▷] key, the symbol → is copied to the cursor location, and the keyboard is set in

ALPHA-lock, which makes each subsequent key press an uppercase alpha character. Alpha characters are printed to the right above the keys.

(2)   Type P M T and then press [ALPHA] to take the keyboard out of ALPHA-lock.

(3)   Press [2nd] [ : ] (the 2nd function of [.]) to begin another command on the same line.

(4)   Press 3 [ × ] 12 [STO ▷] N [ALPHA] to store the expression for the number of periods (years*12) in the variable N. The TI-85 evaluates the expression before stating the value.

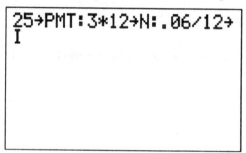

(5)   Press [2nd] [ : ] .06 [ ÷ ] 12 [STO ▷] I [ALPHA] to begin a new command and store the interest per period (rate/12) in the variable I.

The entry is more than 21 characters and cannot be shown in one line of the display, so it "wraps" to the next line.

## 4.4  Entering Expressions

Expressions to be evaluated can contain variable names. On the TI-85, you enter expressions as you would write them on a single line. Enter, for example $PMT((1 + I)\,\hat{}\,(N + 1) - (1 + I))/I$ .

(1)   To enter the expression to define the future value formula, press [2nd] [ : ] to begin the next command press [ALPHA] [ALPHA] to set the keyboard in ALPHA-lock, and then type P M T [ALPHA].

(2)  Press [x] [ ( ] [ ( ] 1 [+] [ALPHA] I [ ) ] [ ^ ] [ ( ] [ALPHA] N [+] 1 [ ) ] [–] [ ( ] 1 [+] [ALPHA] I [ ) ]
     [ ) ] [ ÷ ] [ALPHA] I.

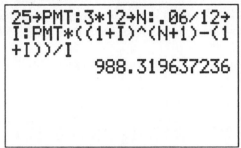

(3)  Press [ENTER] to store the values in the variables and evaluate the expression. The result of the
     expression is shown on the right side of the next line on the display with 12 digits.

(4)  Press [2nd] [MODE] (the 2nd function of [MORE]) to display the mode screen. Press [ ↓ ] [ ▷ ] [ ▷ ]
     [ ▷ ] to position the cursor over the 2 .

(5)  Press [ENTER]. This changes the display format to two fixed decimal places.

(6)  Press [2nd] [QUIT] (the 2nd function of EXIT) which always returns you to the Home screen. Press
     [ENTER]. The last answer is reevaluated and the result is displayed with two fixed decimal places.

     If you save $25 at the beginning of each month for 36 months, invested at 6% , you will have $988.32 .

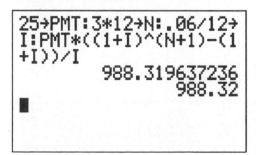

## 4.5  Recalling and Editing a Calculation

If more than one command is entered on a line on the TI-85, the Last Entry feature lets you recall the command that was executed when you pressed [ENTER]. The last result is stored in Last Answer.

   If you continue to invest \$25 a month for another year, how much will you have?

(1)  Press [2nd] [ENTRY]. This recalls the last executed command into the display. The cursor is positioned following the command.

(2)  Use [ ↑ ] and [ ▷ ] to position the cursor over the 3 in the instruction  $3 * 12 \triangleright N$ . Type 4.

```
25→PMT:3*12→N:.06/12→
I:PMT*((1+I)^(N+1)-(1
+I))/I
                988.319637236
                       988.32
25→PMT:4*12→N:.06/12→
I:PMT*((1+I)^(N+1)-(1
+I))/I
```

(3)  You do not need to be at the end of a command to execute it, so press [ENTER] now. The solution for 4 years is displayed on the next line. If you save \$25 at the beginning of each month for 48 months, invested at 6% , you will have \$1359.21.

```
+I))/I
                988.319637236
                       988.32
25→PMT:4*12→N:.06/12→
I:PMT*((1+I)^(N+1)-(1
+I))/I
                      1359.21
```

(4)  If you were able to save \$50 per month, the amount would double because PMT is directly proportional to the total. Press [2nd] [Ans]. The variable name <u>Ans</u> appears in the display.

   Press [ × ] 2 [ENTER].

   You will have \$2718.42 if you save \$50 per month.

```
                       988.32
25→PMT:4*12→N:.06/12→
I:PMT*((1+I)^(N+1)-(1
+I))/I
                      1359.21
Ans*2
                      2718.42
```

## 4.6  Graphing on the TI-85

Users familiar with the TI-81 will find that all of the popular TI-81 graphing features are also on the TI-85. When you press [GRAPH], the menu keys are labeled with the same graphing options (in the same order) that are on the top row of keys on the TI-81.

Graph y=x$^3$-2x and y=2cos x. Determine the solution to x$^3$-2x=2cos x.

(1)  Press [GRAPH]. The menu keys are labeled on the eighth line of the display with the TI-81 **graphing** commands.

The Home screen and cursor are still displayed. You do not leave the Home screen and enter the graphing application until you select a menu key.

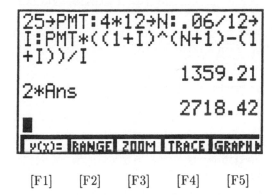

[F1]      [F2]      [F3]      [F4]      [F5]

(2)  Press [F1] to select $< y(x) = >$, which accesses the y(x) editor, where you enter and select functions to graph. Press [x-VAR] (you may press [F1] to select $< x >$ instead) [^] 3 [-] 2 [x-VAR] [ENTER] to enter the equation y1=x^3-2x. Press 2 [COS] [x-VAR] to enter y2=2cos x. The highlighted = shows y1 and y2 are "selected" to be graphed.

Notice, however, that the TI-85 uses lowercase x and y as its graphing variables, rather than the uppercase X and Y used by the TI-81.

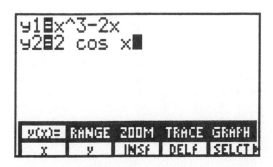

(3)  Press [2nd] and [M3] to select $< ZOOM >$. With the ZOOM instructions, you can easily display the current graph in a different viewing rectangle.

Press [F4] to select < ZSTD >. This is the same as the ZOOM Standard option on the TI-81.

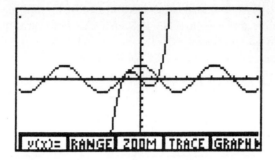

(4)  Press [F4] to select < TRACE >. Press [▶]to trace along function y1, then press [↓] to move to function y2. Notice the 1 or 2 in the upper right of the display, which indicates which function you are tracing.

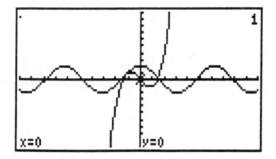

(5)  Press [EXIT] to leave TRACE and display the GRAPH menu.
Press [F3] to select < ZOOM >. Press [F2] to select < ZIN >. Move the cursor over the apparent intersection in the first quadrant. Press [ENTER].

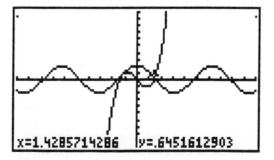

(6)  Press [EXIT] to leave ZIN and display the ZOOM menu.

Press [F4] to select < ZSTD > to display the original graph.

(The coordinate values may vary depending on the cursor location.)

(7)  To explore the apparent solution in the second quadrant, press [F1] to select < BOX >. Move the cursor to the upper right corner of the area you want to examine more closely. Press [ENTER]. Move the cursor to the lower left corner (the box defining the area is shown as you move the cursor). Press [ENTER].

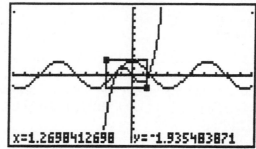

(8)  If necessary, repeat the procedure for ZIN or BOX to see if the two functions intersect in the second quadrant (they do not).

## 4.7  Application: Illumination

On the TI-85, you can explore problems in several different ways  For example, you can solve many problems either by using the <u>Solver</u> feature or graphically. The remaining pages in this chapter present an illumination example to show how to enter equations and explore them both by using the SOLVER and by graphing.

The amount of illumination on a surface is:

(1)  Proportional to the intensity of the source.

(2)  Inversely proportional to the square of the distance.

(3)  Proportional to the sine of the angle between the source and the surface.

The formula for illumination of a point on a surface is:

$$\text{ILLUM} = \frac{\text{INTEN} * \sin\theta}{\text{DIST}^2}$$

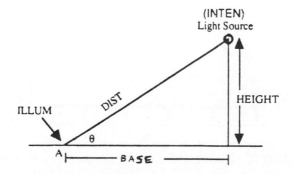

Appropriate units are ft-c (foot-candles) for illumination, CP (candlepower) for intensity, and ft (feet) for distances.

From geometry, $\sin\theta = \dfrac{\text{HEIGHT}}{\text{DIST}}$ .

Therefore, $\text{ILLUM} = \dfrac{\text{INTEN} * \text{HEIGHT}}{\text{DIST}^3}$ .

From geometry, $\text{DIST}^2 = \text{BASE}^2 + \text{HEIGHT}^2$ . On the TI-85, you can store an unevaluated expression as an equation variable.

Press [ALPHA] [ALPHA] to set ALPHA-lock, type DIST=, and then press [ALPHA] to take the keyboard out of ALPHA-lock. Press [2nd] [$\sqrt{\ }$] [ ( ] [ALPHA] [ALPHA] BASE [ALPHA] [x²] [+] [ALPHA] [ALPHA] HEIGHT [ALPHA] [x²] [ ) ] [ENTER].

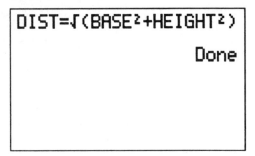

## 4.8  Entering an Equation in the Solver

With the Solver feature of the TI-85, you can solve an equation for any variable in the equation. In the Solver, you can observe the effect that changing the value of one variable has on another and apply "what if" scenarios.

(1)   Press [2nd] [SOLVER] to display the Solver equation entry screen.

(2)   Press [ALPHA] [ALPHA] ILLUM=INTEN [ALPHA] [ × ] [ALPHA] [ALPHA] HEIGHT [ALPHA] [ ÷ ]. Press [F1] to select < DIST > from the menu; the characters DIST are copied to the cursor location.

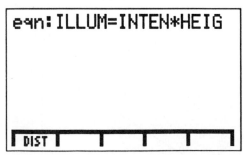

(3)   Press [^]3 to complete the equation that defines illumination in terms of intensity and height: ILLUM = INTEN * HEIGHT/DIST^3. As you enter the equation beyond 17 characters, it scrolls. Ellipsis

marks ( $\cdots$ ) indicate that not all of the equation is displayed on the line. You can use [▶] and [◀] to scroll the equation.

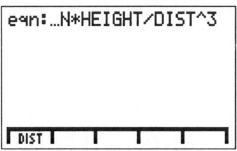

(4)   Press [ENTER]. The Solver edit screen is displayed.

The equation is displayed on the top line. The variables are listed in the order in which they appear in the equation. The variables HEIGHT and BASE, which define the equation variable DIST, are shown. The cursor is positioned after the = following the first variable. If the variables have current values, the value would be shown. Bound defines a limit for the value of the solution. The default values are −1E99 to 1E99.

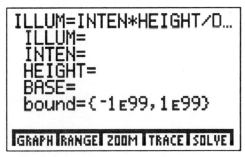

## 4.9  Solving for a Variable

The TI-85 solves the equation for the variable on which the cursor is placed when you select < SOLVE > .

Assume the height of a light on a pole in a parking lot is 50 ft and the intensity of the light is 1000 CP. Determine the illumination on the surface 25 ft from the pole.

(1)   Use [ENTER], [ ↓ ] or [ ↑ ] to move the cursor between the variables. Enter 1000 as the value for INTEN. Enter 50 as the value for HEIGHT. Enter 25 as the value for BASE. The values of INTEN, HEIGHT, and BASE in memory are updated.

(2)  Press [ ↑ ] to move the cursor to <u>ILLUM</u>, the unknown variable. Press [F5] to select < SOLVE > from the menu. A moving bar is shown in the upper right of the display to indicate that the TI-85 is busy calculating or graphing.

The solution is displayed. The square dot next to <u>ILLUM</u> indicates that <u>ILLUM</u> was the variable for which you solved. The value of <u>ILLUM</u> in memory is updated.

<u>lft-rt</u> is the difference between the left side and the right side of the equation, evaluated at the current value of the independent variable.

If the height is 50 ft and the intensity is 1000 CP, the illumination on the surface 25 ft from the pole is .28621670111999 ft-c.

## 4.10  Additional Solutions with the Solver

You can continue to explore solutions to equations with the Solver. You can solve for any variable within the equation to explore "what if" questions.

If the desired illumination is exactly 0.2 ft-c, and the intensity is still 1000 CP, at what height on the pole should the light be placed?

(1)  To change the value of <u>ILLUM</u> to .2, press the [CLEAR] key to clear the value on the line quickly and then type .2. The square dots disappear to show that the solution is not current.

(2)  Move the cursor to <u>HEIGHT</u>. Press [F5] to select < SOLVE >. It is not necessary to clear the value of the variable for which you are solving. If the variable is not cleared, the value is used as the initial guess by the Solver. The equation is solved for <u>HEIGHT</u> and the value displayed.

The illumination on the surface is .2 ft-c and the intensity is 1000 CP, the height of the light source is 63.45876324653 ft.

The solution is dependent on the initial guess and bound.

```
ILLUM=INTEN*HEIGHT/D...
 ILLUM=.2
 INTEN=1000
▪HEIGHT=63.458763246...
 BASE=25
 bound={-1ᴇ99,1ᴇ99}
▪lft-rt=0
┃GRAPH┃RANGE┃ZOOM┃TRACE┃SOLVE┃
```

## 4.11  Changing the Viewing Rectangle

You can graphically examine equations entered in the Solver. The viewing rectangle defines the portion of the graphing coordinate plane that is shown in the display. The values of the RANGE variables determine the size of the viewing rectangle. You can display and edit the values of the RANGE variables.

(1)  Press [F2] to display the RANGE editor.

You display and edit the values of the RANGE variables on this screen. The values shown are the standard default values. The RANGE variables define the viewing rectangle as shown. xMin, xMax, yMin, and yMax define the boundaries of the display. xScl and yScl define the tick marks on the x and y axes.

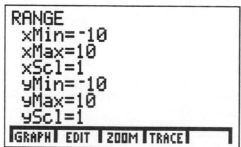

```
RANGE
 xMin=-10
 xMax=10
 xScl=1
 yMin=-10
 yMax=10
 yScl=1
┃GRAPH┃ EDIT ┃ZOOM┃TRACE┃
```

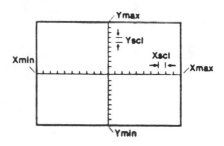

(2)  Graph the illumination example using new values for the RANGE variables, as shown. Use [ ↓ ] or [ENTER] to move the cursor to each value and then type over the existing values to enter the new value. To enter −1 press [(−1)], not [-], and then press 1.

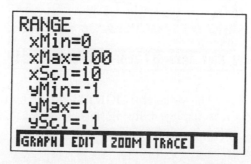

```
RANGE
 xMin=0
 xMax=100
 xScl=10
 yMin=-1
 yMax=1
 yScl=.1
┃GRAPH┃ EDIT ┃ZOOM┃TRACE┃
```

## 4.12  Finding a Solution from a Solver Graph

The graph plots the variable for which the cursor is placed as the independent variable on the x axis and left-rt as the dependent variable on the y axis. Solutions exist for the equation where the function intersects the x axis.

(1)  Press [F1] to select < GRAPH > . The graph plots <u>HEIGHT</u> on the x axis and left-rt on the y axis in the chosen viewing rectangle. The calculation for left-rt in this case is shown below.

$$\underline{\text{lft} - \text{rt}} = \text{ILLUM} - \frac{\text{INTEN} * \text{HEIGHT}}{(\text{BASE}^2 + \text{HEIGHT}^2)^{3/2}}$$

Notice from the graph that this problem has at least two solutions; we found the solution for HEIGHT at the larger value, x=63.458763246529.

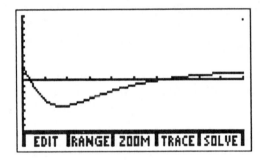

(2)  To solve for the other value of <u>HEIGHT</u>, we must supply a new initial guess or alter the <u>bound</u>. You can select a new initial guess with the graph cursor.

Use [◁] and [▷] to position the cursor near where the function crosses the axis at the smaller value. As you move the cursor, the coordinate values are displayed.

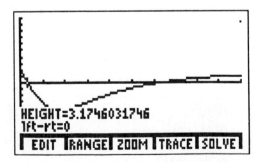

(3)  Press [F5] to select < SOLVE > . The value of HEIGHT identified by the cursor will be used as the new initial guess. The busy indicator is displayed during the calculation. The illumination on the surface is

.2 ft-c and the intensity is 1000 CP, the height of the light source can be either 3.2022212466713 ft or 63.458763246529 ft.

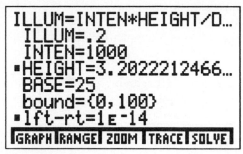

## 4.13 Defining Functions to Graph

On the TI-85, functions are graphed for x and y when x is the independent variable and y=y(x). You can store unevaluated expressions with the = symbol (ALPHA function of the [STO ▷] key). This page shows how to enter the illumination problem for a graphic solution.

Graph the illumination equation and find the height that provides the maximum illumination for a base of 25 feet and an intensity of 1000 CP.

(1)   Press [2nd] [QUIT] to return to the Home screen.

(2)   Press [ALPHA] [ALPHA] HEIGHT= [ALPHA] [x-VAR] [ENTER] to store the unevaluated expression x in an equation variable, HEIGHT. Use [x-VAR] to enter x quickly. INTEN and BASE still contain 1000 and 25.

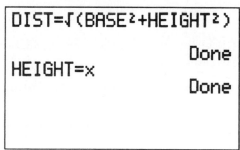

(3)   Press [GRAPH] to display the GRAPH menu. Press [F1] to select < y(x)=> . The display shows the name for the first function, y1.

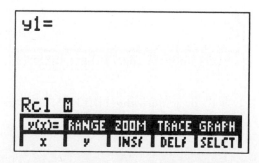

(4)  Press [2nd] [RCL]. The cursor is positioned after Rcl on the sixth line. The RCL feature lets you recall the expression stored in an equation variable to the cursor location. In the Solver, the illumination equation was stored in the equation variable eqn.

(5)  Press [2nd] [ALPHA] to change to lowercase alpha-lock and type eqn [ENTER]. The equation is copied to the cursor location.

(6)  Press [2nd] [◁] to move the cursor to the beginning of the expression quickly. Press [DEL] six times to delete ILLUM=. The highlighted = shows y1 is "selected" to be graphed.

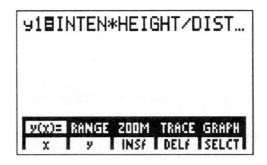

## 4.14  Displaying the Graph

After you have created and selected the function to graph and entered the appropriate viewing rectangle, you can display the graph.

(1)  Press [2nd] [M5] to select < GRAPH > to graph the selected functions in the viewing rectangle. ([2nd] accesses the menu times on the seventh line.)

Because HEIGHT is replaced by x, and the current value of x used, each time a point is plotted. The graph of the function for $0 \leq x \leq 100$ is plotted.

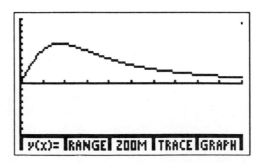

(2)  The graph shows that there is one maximum value of ILLUM for a height between 0 and 100.

Press [ ▷ ] once to display the graphics cursor just to the right of the center of the display. The line above the menu shows the x and y display coordinate values for the cursor position (x,y).

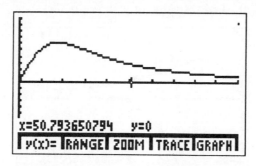

(3)   Using the cursor-movement keys ([ ◁ ], [ ▷ ], [ ↑ ], and [ ↓ ]), move the cursor until it is positioned at the apparent maximum of the function. As you move the cursor, the x and y coordinate values are updated continually with the cursor position.

The free-moving cursor shows maximum illumination of .61290322581 CP for heights from 14.285714286 ft to 21.428271429 ft, within an accuracy of one display dot width. In this example, accuracy$_x$ is .793650793651 and accuracy$_y$ is .032258064516, calculated as shown.

$$\text{Accuracy}_x = \frac{(xMAX - xMin)}{126} = .793650793651 \text{ in this example.}$$

$$\text{Accuracy}_y = \frac{(yMAX - yMin)}{62} = .03225806452 \text{ in this example.}$$

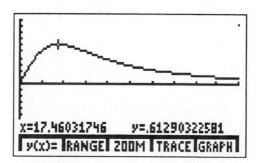

## 4.15  Tracing along a Function

Using the TRACE feature of the TI-85, you can move the cursor along a function, showing the x and y display coordinate values of the cursor location on the function.

(1)   Press [F4] to select < TRACE >. The TRACE cursor appears near the middle of the screen on the function.

The coordinate values of the cursor location $(x,y1(x))$ are displayed on the bottom line of the display. No menu items are shown. The y value shown is the calculated value of the function for the displayed value of x. That is, if $y1=f(x)$, then the value of y shown is $f(x)$.

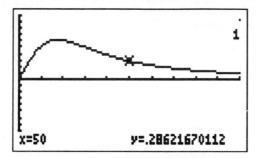

(2)   Use [ ▷ ] and [ ◁ ] to move along a function until you have traced to the largest y value.

The maximum illumination is .61577762623 CP if the height is 17.46031746 ft.

This value of y is the function value $f(x)$ at the x display coordinate value. It is different than the value found with the free-moving cursor, which is based on the RANGE settings.

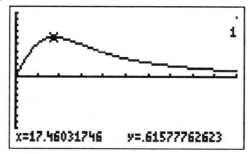

## 4.16  Finding a Maximum Graphically

Within the operations on the GRAPH MATH menu, you can analyze a displayed graph to determine where minimum and maximum values, inflection points, and intercepts occur.

(1)   Press [EXIT] to display the GRAPH menu. Press [MORE] to display additional items on the GRAPH menu.

(2)   Press [F1] to select < MATH >. Press [MORE] to display additional items on the GRAPH MATH menu.

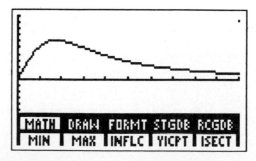

(3)   Press [F2] to select < FMAX >. The TRACE cursor appears near the middle of the screen on the function on the point (x,y1(x)).

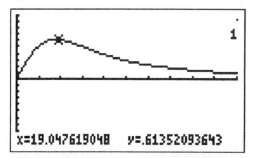

x=19.047619048     y=.61352093643

(4)   Press [ENTER]. The calculator computed maximum is displayed in the cursor coordinates at the bottom of the display, .61584028714 at an x value of 17.677668581. This value of y is larger than the value found with the TRACE cursor. This is the most accurate of the three graphical solutions we have tried. Note: The FMAX and FMIN calculator algorithms search between < LOWER > and < UPPER > and use tol from the [2nd] [TOLER] menu to control accuracy.

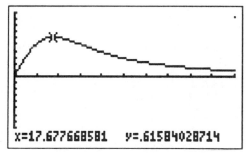

x=17.677668581     y=.61584028714

## 4.17  Graphing the Derivative

The maxima and minima of a continuous differentiable function, if they exist, occur where the first derivative is equal to 0. On the TI-85, you can graph the **exact** derivative of a function.

(1)   Press [GRAPH]. Press [F1] to display the y(x) editor.

Press [ENTER] to move to y2.

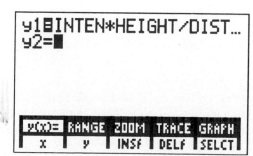

(2)  The calculus functions are grouped on the CALC menu. Press [2nd] [CALC] to display the calculus menu on the bottom line.

(3)  Press [F3]. The function name for the exact first derivative, der1 (, is copied to the cursor location.

(4)  Press [2nd] [M2] to copy y from the menu on the seventh line to the cursor location, then type 1 to enter the name of the first equation, y1. Press [ , ].

(5)  On the TI-85, you can evaluate the calculus functions with respect to any variable, but to be meaningful in graphing, the variable of differentiation or integration should be x.

Press [2nd] [F1] or [2nd][M1] to copy x to the cursor location. Press [ ) ].

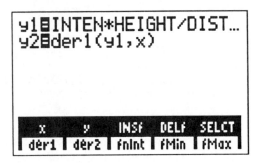

*Remark*   der1(y1,x) is the *exact derivative*, evaluated at the current value of x. When this equation is graphed, the derivative will be calculated for each value of x on the graph.

## 4.18  Zooming In on the Graph

You can magnify the viewing rectangle around a specific cursor location by selecting the Zoom In instruction from the ZOOM menu.

(1)  Press [EXIT] [2nd] [M5] to select < GRAPH > and graph both functions. The busy indicator displays while the graph is plotted. The viewing rectangle is the same as you defined in the Solver, $0 \leq x \leq 100$ and $-1 \leq y \leq 1$. In this viewing rectangle, the graph of the derivative function is very close to the x axis.

(2)   Press [F3] to select  < ZOOM > .

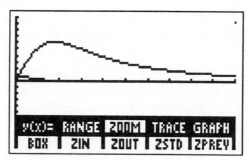

(3)   To zoom in, press [F2] to select  < ZIN >  from the menu.

      The cursor appears at the middle of the display.

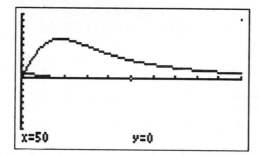

(4)   Use the cursor-movement keys to position the cursor near where the derivative function appears to
      cross the x axis.  Press [ENTER]. The position of the cursor becomes the center of the new viewing
      rectangle. The busy indicator displays while the graph is plotted.

      The new viewing rectangle has been adjusted in both the x and y directions by factors of 4, which are
      the current values for the zoom factors.

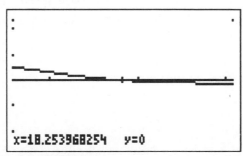

## 4.19  Finding a Root Graphically

The TI-85 can find the root (zero) of a graphed function and can calculate the value of the function for
any value of x. The TI-85 can find the root (zero) of a graphed function and can calculate the value of the

function for any value of x. Find the x value where the root of the derivative function der1(y1,x) occurs and use it to calculate the maximum of the function.

(1)  Press [EXIT] [EXIT] to display the GRAPH menu on the bottom line and press [MORE] to display additional menu items. Press [F1] to select < MATH > to display the GRAPH MATH operations.

(2)  Press [F3] to select < ROOT >. The TRACE cursor is near the middle y value "on" the y1 function, as indicated by the 1 in the upper right corner of the display. The y1 function is "above" the display.

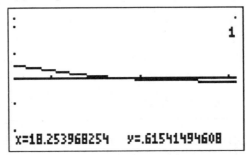

(3)  Press [ ↓ ] to move the cursor to the derivative function, y2, as indicated by the 2 in the upper right corner of the display. You can use [ ▷ ] and [ ◁ ] to move the cursor to a point near the root.

(4)  Press [ENTER]. The busy indicator display while the root is calculated. The calculated root is displayed in the cursor coordinates at the bottom of the display: y=-1.21363E-15 at an x value of 17.67766953.

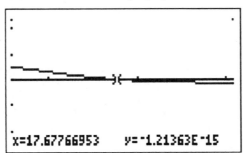

(5)  Press [EXIT] [EXIT] [MORE] [F1] to select < EVAL >. Press [2nd] [Ans] [ENTER] to enter the solution to ROOT as the value for x. The results cursor "displays" on the y1 function at the specified x.

On page 44, < FMAX > found a maximum of y=.61584028714 at x=17.677668581. Corresponding to the maximum, ROOT found a root of the derivative at x=17.67766953, which evaluated to a maximum, y1=.61584028714.

## 4.20  Other Features

This brief chapter introduced you to some basic TI-85 calculator operations, some function graphing features, and one equation solving feature. You should consult the TI-85 User's Guide to learn about these features in more detail and also learn about the many other capabilities of the TI-85. The TI-85 User's Guide Chapter references are noted in parentheses.

### 4.20.1  Some Other Capabilities of the TI-85

(1)  In function graphing, you can store, graph, and analyze up to 99 functions (Chapter 4). In polar graphing, you can store, graph, and analyze up to 99 polar equations (Chapter 5). In parametric graphing, you can store, graph, and analyze up to 99 parametric equations (Chapter 6). In differential equation graphing, you can store, graph, and analyze a differential equation up to nine first-order differential equations (Chapter 7).

(2)  You can use drawing and shading features to add emphasis or perform additional analysis on function, polar, parametric, and differential equation graphs (Chapter 4).

(3)  You can solve an equation for any variable, solve a system of up to 30 simultaneous linear equations, and find the real and complex roots of a polynomial equation (Chapter 14).

(4)  You can enter and store an unlimited number of matrices and vectors with dimensions up to 255. The TI-85 has standard matrix operations, including elementary row operations, and standard vector operations (Chapter 13).

(5)  The TI-85 performs one-variable and two-variable statistical analyses. You can enter and store an unlimited number of data points. Seven regression models are available: linear, logarithmic, exponential, power, and second-, third-, and fourth-order polynomial models. You can analyze data graphically with histograms, scatter plots, and line drawings and plot regression equation graphs (Chapter 15).

(6)  Programming capabilities include extensive control and I/O instructions. You can enter and store an unlimited number of programs (Chapter 16).

(7)  You can share data and programs with another TI-85. You can print graphs and programs, enter programs, and save data on a disk through a PC (Chapter 19).

(8)  The TI-85 has 32K of RAM available to you for storing variables, programs, pictures, and graph databases.

# Chapter 5

**━━━━━━━━━━**

# Calculating and Graphing with
# First Generation Casio Calculators

The purpose of this chapter is to acquaint you with the features of the Casio graphing calculators (models fx-7000G, fx-7500G, fx-8000G, and fx-8500G) that are useful in a graphical approach to precalculus mathematics. Section 5.1 provides some general information about using the Casio and explains how to use its computing and editing features. Section 5.2 discusses the basics of graphing on the Casio without programming. Section 5.3 combines programming with graphics. So get ready to explore and experiment. And always have your Casio handy while reading this chapter.

## 5.1  Numerical Computation and Editing

**5.1.1  The Shift and Alpha Keys:** The distinctively colored $\boxed{\text{SHIFT}}$ and $\boxed{\text{ALPHA}}$ keys are special. Use them as you would a *second function* or *inverse key* on other calculators. That is, any special function or symbol written in gold (blue on the fx-7500G) on the keyboard is accessed by first pressing $\boxed{\text{SHIFT}}$ and then pressing the associated key. Similarly, any alphabetic character or symbol written in red (grey on the fx-7500G) on the keyboard is accessed by pressing $\boxed{\text{ALPHA}}$ followed by the associated key. For example, if you wanted to use the number $\pi$ in your computations, you would key in $\boxed{\text{SHIFT}}$ followed by $\boxed{\text{EXP}}$. In this manual, however, this keying sequence is written $\boxed{\text{SHIFT}}$ $\boxed{\pi}$. To access the letter T, you would press $\boxed{\text{ALPHA}}$ followed by $\boxed{\div}$, but the keying sequence is written $\boxed{\text{ALPHA}}$ $\boxed{\text{T}}$. Once you gain some familiarity with the keyboard, this approach should seem natural.

**A Special Note Concerning (SHIFT).**  The fx-7500G has a different keyboard from the fx-7000G, fx-8000G, and fx-8500G. The fx-7500G has nine more keys than the other models. Nine features accessed by $\boxed{\text{SHIFT}}$ on the fx-7000G, fx-8000G, and fx-8500G do not require a $\boxed{\text{SHIFT}}$ on the fx-7500G. For instance, the exponential function $e^x$ requires no $\boxed{\text{SHIFT}}$ on the fx-7500G. To indicate this, we will write (SHIFT) $\boxed{e^x}$. To access the exponential function on the fx-7500G, you would simply press $\boxed{e^x}$, but on the fx-7000G, fx-8000G, or fx-8500G, you would press $\boxed{\text{SHIFT}}$ $\boxed{e^x}$. Example 5 in Section 5.1.4 illustrates an exponential function computation.

**5.1.2  Getting Ready to Compute:**  Turn on the Casio. The power switch is located on the left side of the fx-7000G and fx-7500G and is located directly below the display screen on the fx-8000G and fx-8500G. If the screen is too light or too dark, adjust it by using the special **contrast** dial on the right edge of the upper keyboard on the fx-7500G. On the fx-7000G, fx-8000G, or fx-8500G, adjust the contrast by first pressing $\boxed{\text{MODE}}$, then pressing $\boxed{\Rightarrow}$ or $\boxed{\Leftarrow}$ several times to darken or lighten the screen as necessary. Your screen should look like Figure 5.1. If it does, go on to Section 5.1.3.

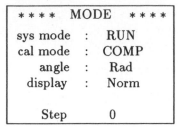

**Figure 5.1.** The screen as it appears when you turn on the Casio.

Whenever you turn on the Casio, the system mode will be RUN. If the calculation mode is not COMP (computation), key in [MODE] [+]. If the angle mode is not Rad (radian), key in [MODE] [5] [EXE]. If the display mode is not Norm (normal), key in [MODE] [9] [EXE]. The modes of the Casio are explained briefly in Sections 5.1.6 and 5.3.1. For details, see your *Owner's Manual.*

### 5.1.3  Error Messages and Editing Expressions

**Cursor/Replay Keys.** The keys [⇐], [⇒], [⇑], and [⇓] on the fx-7000G, fx-8000G, and fx-8500G (or [◀], [▶], [▲], and [▼] on the fx-7500G) are used to move the cursor (blinking ___ ) left, right, up, and down, respectively. These are called *cursor keys*. The keys [⇐] and [⇒] also act as **replay** keys; that is, they return you to the last command statement and allow you to *edit and reexecute* that command.

**Insertions and deletions.** The **insert** feature is used in conjunction with the replay and cursor keys to add one or more symbols to a displayed expression. In general, simply move the cursor to the location of the insertion, and then key in (SHIFT) [INS] followed by one or more symbols. The **delete** key is also used in conjunction with the replay and cursor keys. To delete a character, use the cursor keys to move the cursor to the unwanted symbol and press [DEL].

**Errors and Overwriting.** Sometimes while you are computing or running a program, your Casio will display an **error** message. For example, suppose you wanted to add a positive 7 to a negative 6. Key in [7] [+] [−] [6] [EXE]. The Casio screen should look like Figure 5.2.

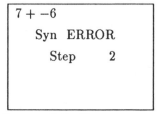

**Figure 5.2.** Error message.

The message in Figure 5.2 means we have made a syntax error at Step 2 (counting begins at Step 0) of our input expression. A syntax error message generally indicates a mistake in a formula or misuse of a program command. Press the [⇐] **replay** key, which takes you immediately to the cause of the error message. At this point, **overwrite** the subtraction operation sign with a negative sign by pressing [(−)]. Then reexecute by pressing [EXE]. You should obtain the expected result, 1.

Now try [9] [÷] [0] [EXE]. You will get a "math" error message. Why?

**5.1.4 Using the "Scientific" Functions:** Like any scientific calculator, the Casio has many built-in functions. But the order in which you press the keys on the Casio differs from most traditional scientific calculators. Here are some examples to get you started using the scientific functions. Once you gain control of these functions, try exploring other keys on your own.

**Powers and Roots.** Press $\boxed{\text{AC}}$ once; then without pressing $\boxed{\text{AC}}$ again, perform the following six computational examples. (Figures 5.3 and 5.4 show how these computations should appear on the Casio screen. Refer to the figures as you proceed through the computations.)

```
(−5)x^y 4
                        625.
4 ^x√2401
                          7.
sin 45°
             0.7071067812
√2 ÷ 2
             0.7071067812
```

**Figure 5.3.** Power, root, and trigonometric computations.

```
eπ
               23.14069263
Int −3.749
                        −3.
5 → A
                         5.
4A² − 3 → R
                        97.
```

**Figure 5.4.** Assorted computations.

(1) Evaluate $(-5)^4$ using $\boxed{(}$ $\boxed{(-)}$ $\boxed{5}$ $\boxed{)}$ $\boxed{x^y}$ $\boxed{4}$ $\boxed{\text{EXE}}$. Remember not to confuse the additive inverse, or "sign change," key $\boxed{(-)}$ with the subtraction key $\boxed{-}$.

(2) Simplify $\sqrt[4]{2401}$ with the keying sequence $\boxed{4}$ $\boxed{^x\sqrt{}}$ $\boxed{2401}$ $\boxed{\text{EXE}}$.

(3) To calculate the sine of a 45-degree angle without switching to degree mode, key in $\boxed{\text{sin}}$ $\boxed{45}$ $\boxed{\text{SHIFT}}$ $\boxed{\text{MODE}}$ $\boxed{4}$ $\boxed{\text{EXE}}$.

(4) In trigonometry you learn that $\sin 45° = \frac{\sqrt{2}}{2}$. Press $\boxed{\sqrt{}}$ $\boxed{2}$ $\boxed{÷}$ $\boxed{2}$ $\boxed{\text{EXE}}$. Look at the answer, and compare it with the previous answer on the screen display.

(5) Exponential function computations look strange on the Casio. For example, when you key in (SHIFT) $\boxed{e^x}$) $\boxed{\text{SHIFT}}$ $\boxed{\pi}$ $\boxed{\text{EXE}}$, the screen display suggests that $e$ has been multiplied by $\pi$, but actually $e$ has been raised to the power $\pi$ (See Figure 5.4).

(6) The accompanying textbook discusses the **greatest integer function** INT. It is easy to confuse the greatest integer function with the Casio **integer truncation function** Int. To see that they are different functions, press $\boxed{\text{SHIFT}}$ $\boxed{\text{Int}}$ $\boxed{(-)}$ $\boxed{3.749}$ $\boxed{\text{EXE}}$. On the Casio, Int(-3.749) = -3, but the

greatest integer less than or equal to -3.749 is -4; so, INT(-3.749) = -4. The moral to this story is: **There is no direct way to access the greatest integer function on the Casio.**

**5.1.5   Memory and the Assignment Arrow:** The uppercase letters A through Z on the Casio keyboard act as the names for 26 memory locations, or **storage registers**. Use the assignment key $\rightarrow$ to assign a numerical value to a storage register. (Don't confuse the assignment key $\rightarrow$ with the cursor key $\Rightarrow$.) For example, key in the sequence $\boxed{5}$ $\boxed{\rightarrow}$ $\boxed{\text{ALPHA}}$ $\boxed{\text{A}}$ $\boxed{\text{EXE}}$ to assign the number 5 to storage register A. The value **5.** should appear on the right side of the display screen (see Figure 5.4). Key in $\boxed{4}$ $\boxed{\text{ALPHA}}$ $\boxed{\text{A}}$ $\boxed{x^2}$ $\boxed{-}$ $\boxed{3}$ $\boxed{\rightarrow}$ $\boxed{\text{ALPHA}}$ $\boxed{\text{R}}$ $\boxed{\text{EXE}}$ to compute $4A^2 - 3$ and assign the value to R. Notice that the Casio understands **left juxtaposition** as multiplication, that is, it multiplies the 4 by the $A^2$.

A value stored in a register will remain there until you assign the register a different value, *even if you turn off the machine.* To check this, turn the calculator off, then on again, and key in $\boxed{\text{ALPHA}}$ $\boxed{\text{A}}$ $\boxed{\text{EXE}}$. The value **5.** should appear at the right of the screen (see Figure 5.5). The keying sequence $\boxed{\text{ALPHA}}$ $\boxed{\text{R}}$ $\boxed{\text{EXE}}$ will reveal the contents of R.

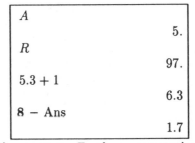

**Figure 5.5.** Further computations.

**The Answer Key.** The Casio has a special storage register in which the answer to the *last* computation is automatically stored. To check its contents, key in $\boxed{\text{Ans}}$ $\boxed{\text{EXE}}$. The stored value can be used in further computations. As a simple example, key in $\boxed{5.3}$ $\boxed{+}$ $\boxed{1}$ $\boxed{\text{EXE}}$ (see Figure 5.5). The value **6.3** is now stored in the "Ans" memory. Now key in $\boxed{8}$ $\boxed{-}$ $\boxed{\text{Ans}}$ $\boxed{\text{EXE}}$. The answer **1.7** should appear on the display screen (and is automatically placed in Ans).

**Input Buffer.** On the fx-8000G and fx-8500G, $\boxed{\text{MODE}}$ $\boxed{\text{Ans}}$ acts as an input buffer recall. Similar to replay, it recalls the previous input even if you press $\boxed{\text{AC}}$ or turn off the machine. If you're using one of these models, try it! The fx-7000G and fx-7500G have no input buffers.

**5.1.6  Miscellaneous Matters**

**Modes.** The Casio has many capabilities. For each particular use, the status of the unit must be set appropriately. When you first turn on your calculator, you see the **mode display window**. At other times, you can check to see which modes are active by pressing $\boxed{\text{M}\text{-Disp}}$, but the mode display window will remain on the screen only while this key is depressed. Generally, we will operate in the modes shown in Figure 5.1. *You are encouraged to confirm which modes are active before calculating, graphing, or programming on the Casio.*

The mode-setting codes are shown on your calculator directly below the display screen on the fx-7000G, fx-8000G, and fx-8500G and above the lower keyboard on the fx-7500G. Modes 1, 2, and 3 are used to run, write, or delete programs. These are explained in Section 5.3.1. Modes 4, 5, and 6 are used to choose degree, radian, or grad **angle** measure. Modes 7, 8, and 9 specify a fixed-decimal-place, scientific notation, or "normal" **display**. The **calculation** mode should be set to COMP using $\boxed{\texttt{MODE}}$ $\boxed{+}$, unless you wish to do statistical computations or operate in bases other than base 10. For full details, consult your *Owner's Manual*.

**Display Windows.** In all, the Casio features four display windows:

(1)   The mode display window.

(2)   The text display window.

(3)   The graphics display window.

(4)   The Range settings display window.

Figure 5.1 shows an example of the mode display window. Most of the time while working through this section your calculator was displaying its text window. Calculating, table building, and program writing all occur in the text display window. In Section 5.2 you will be introduced to the Range settings display window, the graphics display window, and to the $\boxed{\texttt{Range}}$ and $\boxed{\texttt{G} \leftrightarrow \texttt{T}}$ "toggle" keys that allow you to move in and out of these two windows.

**Automatic Power Off.** The Casio's power is automatically switched off approximately 6 minutes after the last operation. The display screen goes blank, and you lose any formulas from the text screen, but the memory, programs, Range, and graphics-window contents are all retained. You can restore the power by either switching the machine off, then on, or by pressing $\boxed{\texttt{AC}}$.

## 5.2   Casio Graphing Fundamentals

Set your system, calculation, and angle modes to RUN, COMP, and RAD (for help, see Section 5.1.2). These modes are used throughout this chapter.

In this section we will explore graphing techniques that require no programming. The key to any computer-based graphing is learning how to control and select viewing rectangles so that you can see the parts of the graph or graphs you are interested in. On the Casio the **Range** feature is used to set the viewing rectangle as well as the scale marks on each axis. Choosing and changing viewing rectangles is a major focus of this section.

**Ways to Change the Viewing Rectangle.** There are six ways to change from one viewing rectangle to another on the Casio. These are listed below together with the section where each method is introduced.

(1)   Keying in Range settings (Section 5.2.1).

(2)   Graphing a built-in function without using X (Section 5.2.1).

(3)   Pressing $\boxed{\texttt{SHIFT}}$ $\boxed{\texttt{Mcl}}$ while in the Range settings window (Section 5.2.1).

(4)   Using automatic zoom-in, $\boxed{\texttt{SHIFT}}$ $\boxed{\times}$, or zoom-out, $\boxed{\texttt{SHIFT}}$ $\boxed{\div}$ (Section 5.2.2).

(5)   Using the Factor feature to zoom-in or zoom-out (Section 5.3.4).

(6)   Setting the Range within a program (Programming Hint 2, Section 5.3.3). The Zoom-In Program given in Section 5.3.6 is an especially useful example of this.

The abilities to **clear the graphics window** and to **overlay graphs** are also important to graphical problem solving. These are explained in Sections 5.2.1 and 5.2.2, respectively.

### 5.2.1 Graphing and Range Settings

**Graphing Built-in Functions.** Key in $\boxed{\text{Graph}}$ $\boxed{x^2}$ $\boxed{\text{EXE}}$. Figure 5.6 shows a pixel for pixel facsimile of the Casio graph that is produced. After the graph is drawn, press the $\boxed{\text{Range}}$ key. The Range setting values should be:

$$\text{Xmin} = \text{-7} \qquad \text{Xmax} = 7 \qquad \text{Xscale} = 2 \qquad \text{Ymin} = \text{-2} \qquad \text{Ymax} = 29 \qquad \text{Yscale} = 5$$

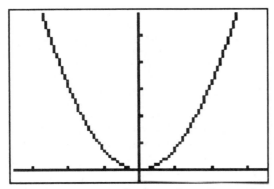  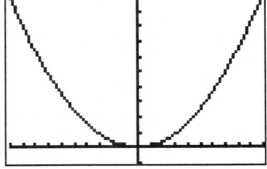

**Figure 5.6.** $y = x^2$, $[-7, 7]$ by $[-2, 29]$.     **Figure 5.7.** $y = \sin x$, $[-2\pi, 2\pi]$ by $[-1.6, 1.6]$.

Press $\boxed{\text{Range}}$ again to toggle back to the graphics window, then key in $\boxed{\text{G} \leftrightarrow \text{T}}$ to return to the text window. The key $\boxed{\text{G} \leftrightarrow \text{T}}$ allows you to alternate (toggle) between the graphics and text windows. Key in $\boxed{\text{Graph}}$ $\boxed{\text{sin}}$ $\boxed{\text{EXE}}$ (see Figure 5.7). What are the Range values now? Have they changed? They should have. They should now be essentially the following values. (The Casio actually uses and displays decimal approximations of the Xmin, Xmax, and Xscale values listed below.)

$$\text{Xmin} = \text{-2}\,\pi \qquad \text{Xmax} = 2\,\pi \qquad \text{Xscale} = \pi \qquad \text{Ymin} = \text{-1.6} \qquad \text{Ymax} = 1.6 \qquad \text{Yscale} = 0.5$$

Graph some other built-in functions, that is, those accessible using $\boxed{\text{Graph}}$ $\boxed{\text{key}}$ $\boxed{\text{EXE}}$ or $\boxed{\text{Graph}}$ $\boxed{\text{SHIFT}}$ $\boxed{\text{key}}$ $\boxed{\text{EXE}}$. Each built-in function on the Casio has built-in Range settings. Try graphing $\log x$, $\ln x$, $e^x$, $\cos x$, $\tan x$, $x^{-1}$, and other built-in functions. Check the Range values after executing each graph.

**Clearing the Graphics Window.** Whenever any of the Range settings are changed, the graphics display window is automatically cleared. At times you may wish to clear all graphs from the graphics window without changing the viewing rectangle or scale marks. To do this, press (SHIFT) $\boxed{\text{Cls}}$ $\boxed{\text{EXE}}$. You can check to see that the graphics window has been cleared by pressing $\boxed{\text{G} \leftrightarrow \text{T}}$. The graphs are removed, but the scaled axes are not.

**Choosing the Viewing Rectangle You Want.** You can select any Range values you wish, *but then you must provide the function with the argument X*. If the Range settings window is not already showing on your display screen, press $\boxed{\text{Range}}$. Then key in the values

$$\text{Xmin} = \text{-10} \qquad \text{Xmax} = 10 \qquad \text{Xscale} = 1 \qquad \text{Ymin} = \text{-10} \qquad \text{Ymax} = 100 \qquad \text{Yscale} = 10$$

pressing $\boxed{\text{EXE}}$ after keying in each of the six values. These Range values are referred to as the **viewing rectangle** $[-10, 10]$ by $[-10, 100]$ with Xscale $= 1$ and Yscale $= 10$.

**Agreement.** In this manual, Range values are specified in terms of viewing rectangles, and you are left to choose appropriate Xscale and Yscale values.

   Now key in $\boxed{\text{Graph}}$ $\boxed{\text{ALPHA}}$ $\boxed{\text{X}}$ $\boxed{x^2}$ $\boxed{\text{EXE}}$. Check the Range (see Figure 5.8).

**Setting the Casio Default Range.** If the Range settings window is not showing, press $\boxed{\text{Range}}$. Then key in $\boxed{\text{SHIFT}}$ $\boxed{\text{Mcl}}$ to enter the **Casio default viewing rectangle** of $[-4.7, 4.7]$ by $[-3.1, 3.1]$. (Note: As described in Section 2.3, this choice yields a $\Delta x$ of 0.1 (and also a corresponding change in $y$, $\Delta y = (\text{Ymax} - \text{Ymin})/62$, of 0.1).) Notice that the Casio default viewing rectangle is different from the **standard viewing rectangle** of $[-10, 10]$ by $[-10, 10]$ used in the accompanying textbook. Press $\boxed{\text{Range}}$ to get out of the Range window, and then key in $\boxed{\text{Graph}}$ $\boxed{\text{sin}}$ $\boxed{\text{EXE}}$. After the graph is drawn press $\boxed{\text{Range}}$ (see Figure 5.8). What happened? The Casio automatically uses the built-in Range for each function if you leave out the X. Now reset the default values by keying in $\boxed{\text{SHIFT}}$ $\boxed{\text{Mcl}}$ $\boxed{\text{Range}}$ again, and then press $\boxed{\text{Graph}}$ $\boxed{\text{sin}}$ $\boxed{\text{ALPHA}}$ $\boxed{\text{X}}$ $\boxed{\text{EXE}}$. Check the Range (see Figure 5.9).

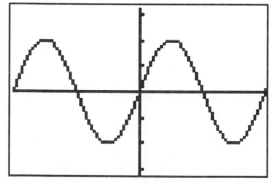

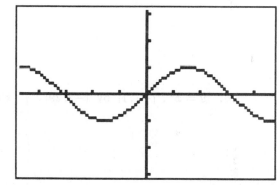

**Figure 5.8.** $y = x^2$, $[-10, 10]$ by $[-10, 100]$.   **Figure 5.9.** $y = \sin x$, $[-4.7, 4.7]$ by $[-3.1, 3.1]$.

**5.2.2   Overlaying Graphs, Trace, and Automatic Zooming:**   Now that you are familiar with how to set Range values, the next steps in graphing on the Casio are to overlay graphs and to use the Trace and automatic zoom features. These steps are demonstrated in the equation solving problem given below.

**Overlaying Graphs.** The $\boxed{:}$ key can be used to string together a series of command statements. This method can be used to graph several functions—one after the other in the same viewing rectangle—by stringing together several Graph statements.

**Trace.** To locate points on a graph, we use the Trace feature. This feature allows dot-to-dot movement along the most recently drawn graph, with the calculator displaying the $x$- or $y$-coordinate associated with each dot, or **pixel**, along the way. It is easy to stop, reverse direction, and switch from an $x$-readout to a $y$-readout or vice versa.

**Automatic Zoom.** The Casio *Owner's Manual* calls this the "instant factor function." Early versions of the fx-7000G did not have this feature. Now all four Casio graphing calculator models possess this important capability. You can either **zoom-in** to study a small portion of a graph or graphs or **zoom-out** to investigate global behavior.

Using these Casio features you can graphically solve equations, inequalities, and systems of equations. These features open the door to solving extreme-value (max/min) problems and to determining intervals over which a function is increasing or decreasing. The same methods apply regardless of the functions involved. The equation $\cos x = \tan x$ can be solved by the same graphical method that would be used to solve $2x = 6$.

*Problem*   Solve $\cos x = \tan x$ for $0 \le x \le 1$.

*Solution*   Clear the graphics screen and reset the default viewing rectangle by keying in $\boxed{\text{Range}}$ $\boxed{\text{SHIFT}}$ $\boxed{\text{Mcl}}$ $\boxed{\text{Range}}$. Then enter the functions by pressing $\boxed{\text{Graph}}$ $\boxed{\cos}$ $\boxed{\text{ALPHA}}$ $\boxed{\text{X}}$ $\boxed{:}$ $\boxed{\text{Graph}}$ $\boxed{\tan}$ $\boxed{\text{ALPHA}}$ $\boxed{\text{X}}$. The text screen should look like this:

$$\text{Graph Y} = \cos \text{X} : \text{Gr}$$

$$\text{aph Y} = \tan \text{X}$$

Edit if necessary, and then press $\boxed{\text{EXE}}$. Watch carefully: Notice that the cosine function is drawn first, and then the tangent function is **overlaid** (see Figure 5.10).

You can approximate the coordinates of any point on the most recently graphed function with the **Trace** feature. Press (SHIFT) $\boxed{\text{Trace}}$. Locate the blinking pixel at the left of the screen. Use the right cursor key $\boxed{\Rightarrow}$ repeatedly to move the blinking pixel to the point of intersection of the graphs of $y = \cos x$ and $y = \tan x$ that lies between $x = 0$ and $x = 1$. The $x$-coordinate should be $0.7$ (see Figure 5.11). Now press $\boxed{\text{SHIFT}}$ $\boxed{\text{X} \leftrightarrow \text{Y}}$. What is the $y$-coordinate?

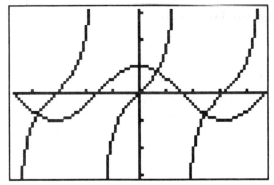 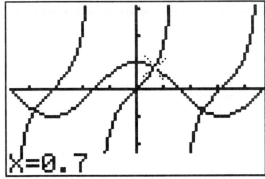

**Figure 5.10.**   Cosine and tangent overlaid.      **Figure 5.11.**   Trace approximating intersection.

**Automatic Zoom-In.**   With the picture in Figure 5.11 showing on the screen, press $\boxed{\text{SHIFT}}$ $\boxed{\times}$. Notice that the Casio uses the traced-to point as the center of the new, smaller viewing rectangle. The automatic zoom factor is 2 in both the $x$ and $y$ directions. Press $\boxed{\text{SHIFT}}$ $\boxed{\times}$ again. If you did not use Trace, the Casio zooms in about the center of the current viewing rectangle. Without using Trace and waiting for the new graphs each time, zoom in three more times (for a total of 5 zoom-in steps). You should obtain the view shown in Figure 5.12.

Now use Trace again to locate the new, improved approximation for the $x$-coordinate of the point of intersection (see Figure 5.13).

You could follow this by repeated use of $\boxed{\text{SHIFT}}$ $\boxed{\times}$ to obtain further accuracy.

**Automatic Zoom-Out.**   To zoom-out automatically (on all but early versions of the fx-7000G), press $\boxed{\text{SHIFT}}$ $\boxed{\div}$.

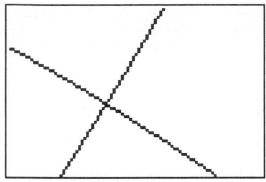

**Figure 5.12.** After five zoom-in steps.

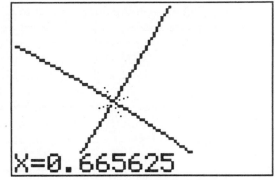

**Figure 5.13.** Trace used again.

**Alternative Zooming Methods.** When you first begin using the Casio, you may be more comfortable zooming-in and zooming-out by changing the Range settings by hand. Eventually you will probably use a variety of methods depending on the situation. Section 5.3.4 explains how to use the Factor feature in conjunction with stored graphing statements—an approach that is more versatile than automatic zoom and that is available on all Casio machines, even old fx-7000Gs. For zooming-in on any Casio, the Zoom-In Program of Section 5.3.6 is fast, effective, and visually appealing.

## 5.3 Combining Graphics with Programming on the Casio

Sections 5.3.1 through 5.3.3 provide an introduction to Casio programming. You may wish to read these sections quickly now, and then refer to them later as questions arise. Section 5.3.4 shows how to graph from a program, which can save time and avoid the frustration of having to retype a function to be graphed. Section 5.3.5 discusses the Plot and Line features that are put to use in Sections 5.3.6–5.3.8 in useful graphing programs.

**5.3.1 Programming Modes:** The Casio computer has three modes of operation related to programming: Modes 1, 2, and 3. **Mode 1** is the **RUN** mode. The computer must be in this mode to run stored programs. The Mode 1 screen is the screen that first appears when the computer is turned on. Figure 5.1 shows this screen. Notice that the first line of information indicates the system is in RUN mode.

**Mode 2** is the **WRT** mode, or the program writing mode. This is the mode used to write, edit, and store programs. There are 10 program storage locations numbered **0** through **9**. These 10 storage locations are represented on the Mode 2 screen by the 10 digits at the bottom of the screen (see Figure 5.14).

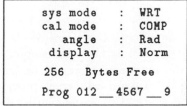

```
sys mode   :  WRT
cal mode   :  COMP
   angle   :  Rad
 display   :  Norm

256  Bytes Free

Prog 012 __ 4567 __ 9
```

**Figure 5.14.** Mode 2 screen; system in WRT mode.

Empty program storage locations are indicated by a visible digit, and locations containing programs are indicated by a "__" in place of the digit. In Figure 5.14, program storage locations 3 and 8 contain programs, while the remaining locations are empty. To select a program storage location, use the $\boxed{\Leftarrow}$ and $\boxed{\Rightarrow}$ cursor keys to move through the 10 digits. The location selected will flash. To view an existing program or to write a new program press $\boxed{\texttt{EXE}}$ when the location of your choice is flashing. If you pick a location with an existing program, the program is displayed on the screen. If you pick an empty location, a blank screen with the cursor in the upper left corner should appear, indicating a new program may be entered. As statements are keyed in, they are displayed on the screen and automatically saved in the memory of the computer. Even if you were to switch the computer off in the middle of a program, all the work done up to that point will be saved. Once a new program is written or an existing program is edited, the computer must be returned to Mode 1 to run the program. This is done by keying $\boxed{\texttt{MODE}}$ $\boxed{\texttt{1}}$.

Mode 3 is the PCL mode, or the program clearing mode. This mode is used to erase existing programs from the memory of the computer. The Mode 3 screen is identical to the Mode 2 screen except that the system mode is PCL (see Figure 5.15).

```
┌─────────────────────────────┐
│   sys mode   :  PCL          │
│   cal mode   :  COMP         │
│     angle    :  Rad          │
│   display    :  Norm         │
│                              │
│   256   Bytes Free           │
│                              │
│   Prog 012__4567__9          │
└─────────────────────────────┘
```

**Figure 5.15.** Mode 3 screen; system in PCL mode.

**Caution should be exercised when using Mode 3** because you may mistakenly delete a program you wish to save. Selection of program storage locations is the same in Mode 3 as in Mode 2. To delete a single program, select the storage location using the $\boxed{\Leftarrow}$ and $\boxed{\Rightarrow}$ cursor keys and press $\boxed{\texttt{AC}}$. After a slight delay the number of that location will appear indicating an empty storage location. To clear all 10 programs at once, press $\boxed{\texttt{SHIFT}}$ $\boxed{\texttt{Mcl}}$. All of the digits 0 through 9 will then appear. Remember, once a program storage location is cleared, the stored program is gone. There is no temporary storage buffer to give you a second chance. When you are finished erasing programs, you may return either to Mode 2 to write programs or to Mode 1 to run programs.

### 5.3.2 Writing, Running, and Editing Programs

**Writing a Program.** To write a program, enter Mode 2 by keying $\boxed{\texttt{MODE}}$ $\boxed{\texttt{2}}$, select an empty program storage location and press $\boxed{\texttt{EXE}}$, and then type the code (key in the statements) of the program. Anything you type will be saved. When you have finished typing the program, exit the writing mode and return to the RUN mode by keying $\boxed{\texttt{MODE}}$ $\boxed{\texttt{1}}$.

**Running a Program.** Programs can be run only from Mode 1. To run a program put the computer in Mode 1 by keying $\boxed{\texttt{MODE}}$ $\boxed{\texttt{1}}$. This is also the default system mode when the computer is turned on. Programs are stored in the numbered storage locations 0 through 9. To run a program, say Program 4, key in $\boxed{\texttt{Prog}}$ $\boxed{\texttt{4}}$ $\boxed{\texttt{EXE}}$. To run a different program use the number of that program in the previous keying sequence. This method of naming programs by number means that you must remember which program is stored in which location. When a program is finished running, the computer is still in the RUN mode; no special commands are necessary to continue. To run the same program again, press $\boxed{\text{EXE}}$; the last program command statement will be reexecuted, running the program again.

**Editing a Program.** Program editing is done just like writing a program. Enter Mode 2 by keying $\boxed{\text{MODE}}$ $\boxed{2}$; select the location of the program you wish to edit; and press $\boxed{\text{EXE}}$. The program will be shown on the screen. Use the cursor arrow keys ($\boxed{\Leftarrow}$, $\boxed{\Rightarrow}$, $\boxed{\Uparrow}$, and $\boxed{\Downarrow}$) to move the cursor to the point in the program you wish to edit. Use the Insert and Delete features to insert or remove symbols or entire commands. When the editing is finished, return to Mode 1 to run the program. No special commands are needed to save the changes in the program.

### 5.3.3 Programming Hints

(1) **The uses of EXE and ◢ in programming.** When entering the programs listed in this manual, each line in a program listing should be finished with an $\boxed{\text{EXE}}$ (Execute) keystroke. It appears that nothing is entered into the program, but the cursor moves to the next full line. This keystroke operates like a carriage **return** when typing. The EXE key enters an invisible character that separates command lines in a program.

Programs listed in your Casio *Owner's Manual* do not use EXE commands to end lines, but rather **concatenate** (string together) the entire program with colons (:) between each command statement. This makes programs very difficult to read. Using EXE commands at the end of each line makes program listings easier to read and edit, and takes no more memory than using the colons to concatenate commands.

When there are exactly 16 characters in a command line of a program, it appears as if the cursor has moved to a new program line. You still must enter an EXE character. The program seems to have an extra line. *It does not.* However, you should never press $\boxed{\text{EXE}}$ twice at the end of a line; if you do, a syntax error message will occur when you try to run the program.

Program editing is done using the INS (insert) and DEL (delete) features. When inserting an EXE command, it is necessary to use the INS feature even though no character appears on the screen. Placing the cursor at the end of a program line and pressing $\boxed{\text{DEL}}$ will delete an invisible EXE character.

The ◢ character at the end of a command line acts like an EXE keystroke. Once entered, no EXE keystroke is needed to end the line; the cursor automatically moves to the beginning of the next programming line. This command causes the program to output a numerical calculation and display the **-Disp-** message on the screen. Without this command, intermediate calculations will *not* be printed on the screen. The *last* calculation in a program will be printed on the screen *without* using this command.

(2) **Setting the Range within a program.** The Range of the graphics screen may be set from within a program. This is done by entering the six Range values in order separated by commas. For example, the command line

Range −10, 30, 5, −1000, 3500, 500

would set a viewing rectangle of $[-10, 30]$ by $[-1000, 3500]$. The horizontal scale marks would represent 5 units, and the vertical scale marks would represent 500 units. Range values may also be designated by variable quantities, as illustrated in the program given within the next paragraph.

Recall that the viewing rectangle $[-6.7, 6.7]$ by $[-3.1, 3.1]$ is the Casio default viewing rectangle, which can be set using $\boxed{\text{SHIFT}}$ $\boxed{\text{Mcl}}$ while the Range display window is active. These default Range values make each pixel on the screen represent $0.1$ in both the horizontal and vertical directions. In effect they put the graphics display in "square" coordinates. When plotting shapes that should look a certain way, such as circles or perpendicular lines, this square coordinate system gives an accurate visual representation of the shapes. Multiples of the default scale also preserve the shape of pictures. The Range settings given above generate any multiple of the default viewing rectangle quickly and accurately. Using the values $1, 2, 5$, or $10$ for F makes each pixel a step of $0.1, 0.2, 0.5$, or $1$, respectively. These steps are useful when doing analytic geometry problems, and they give nice Trace coordinate readouts. The following simple program can be used to set the Range to any multiple of the default viewing rectangle.

$$\text{``SCALE FACTOR'' ?} \rightarrow F$$
$$\text{Range } -4.7F, \ 4.7F, \ F, \ -3.1F, \ 3.1F, \ F$$

When you run this program the words SCALE FACTOR will appear on the screen, followed by a question mark, or **input prompt symbol**. Words written inside quotation marks within a program will always appear as screen messages when the program is run, and the input prompt symbol (?) is used in conjunction with the assignment arrow ($\rightarrow$) to permit the assignment of values to variables while running a program (extending the methods of Section 5.1.5). To respond to the screen message in this case, key in the value you wish to use for F, and then press the EXE key. This will set the Range as desired.

(3)   **Error messages.** When you have an error in a program, the Casio will stop the program and print a message indicating the type of error, the program containing the error, and the step within the program causing the error. For example, the error message

$$\text{Syn ERROR}$$
$$\text{Step P9 } - 65$$

would indicate a syntax error at Step 65 of Program 9. Returning to this point in Program 9 is done by using a REPLAY key ($\boxed{\Leftarrow}$ or $\boxed{\Rightarrow}$). After the error is corrected, return to Mode 1, and run the program again.

Command words in the programs that appear in upper and lower case (e.g., Goto) are reserved words in the Casio programming language. These commands are accessed by single keystrokes. Do not type these words using the alphabetic keys. For example, typing the four letters **GOTO** instead of the single command **Goto** will generate an error message.

(4)   **Program storage locations.** Programs may be stored in any of the 10 storage locations you wish. However, some of the programs in this manual use other programs as subroutines to do specific tasks. When these subprograms are used, you must be certain that the correct subprogram is called from the main program. For example, the Zoom-In Program in Section 5.3.6 calls Program 0 as a subprogram. So, the functions to be graphed should be stored in Program 0. If those functions were stored in Program 3, then the command line in the Zoom-In Program would have to call Program 3 to find the functions. You may customize your programs any way you wish.

**5.3.4   Graphing from a Program:** One of the most convenient uses of programming on the Casio is to graph functions from a program. Often beginners become frustrated when attempting to manipulate a graph without programming. Unintended or mistaken keystrokes can cause the loss of the command lines containing the functions to be graphed. This often occurs when using the Trace and Factor features to zoom-in on a graph.

To eliminate this problem, enter the functions to be graphed as a program. Then to graph the functions, simply run the program. Let's look at an example.

*Problem*   Solve the system of equations $y = \cos x$ and $y = x^3 - 1.4x$.

*Solution*   Enter Mode 2 and select program storage location 0. (This is a handy location to use for graphing functions because it is the location selected automatically when Mode 2 is activated.) Enter the following keying sequence (see Figure 5.16).

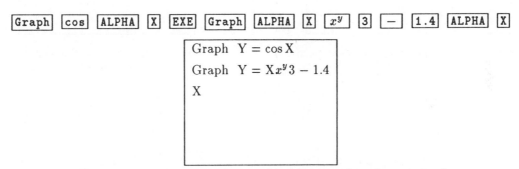

**Figure 5.16.** Program for graphing the two functions entered.

Notice that the EXE key was pressed after the first line was entered, but not after the last. When writing programs, the EXE key acts like the return key on a computer. Pressing EXE writes an invisible character at the end of a command line that separates it from the subsequent line. You need not use an EXE at the end of a program; more than one EXE at the end of a program will cause the Casio to display an error message (see Programming Hint 1 in Section 5.3.3).

Once the program is written, return to Mode 1 (key MODE 1) to run the program. Set the Range to the default viewing rectangle of $[-4.7, 4.7]$ by $[-3.1, 3.1]$ by keying Range SHIFT Mcl Range. Now, graph the two functions by keying Prog 0 EXE. Figure 5.17 resembles the screen after the two functions have been graphed. (The graphic figures in this section were produced using *Master Grapher*, Waits & Demana, 1989, and are not exact facsimiles of Casio graphs.)

At this point, the program has finished running, and all the built-in features used to manipulate graphs are available. For example, pressing SHIFT × will result in a zoom-in by a factor of 2 in the horizontal and vertical directions. Pressing SHIFT ÷ will cause a zoom-out by a factor of 2 in both directions. The Trace feature can be activated by keying (SHIFT) Trace, and then the right and left cursor keys can be used to move the blinking dot along the last function graphed. The $x$- and $y$-coordinates of the blinking point are given at the bottom of the screen. Key SHIFT X↔Y to toggle between the $x$ and $y$ values.

If you happen to press the wrong key, the functions can be regraphed by running Program 0 again with the keystrokes Prog 0 EXE.

*Problem extension*   Use the Factor feature to zoom-in on the solution(s) to the system of equations $y = \cos x$ and $y = x^3 - 1.4x$.

*Solution*   After running Program 0 to graph the functions, use the Trace feature to place the blinking cursor on (or as close as possible to) the intersection of the two functions in the second quadrant (see Figure 5.18).

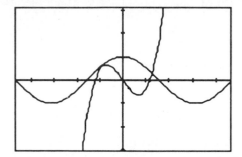

**Figure 5.17.**  Functions graphed from Program 0.

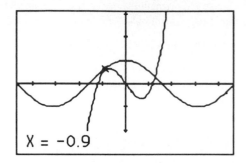

**Figure 5.18.**  Trace in the second quadrant.

Next, enter the keystrokes (SHIFT) `Factor` `5` `:` `Prog` `0` `EXE`. This will cause a zoom-in by a factor of 5 in both directions. Figure 5.19 shows the graphs after this zoom-in. It is clear now that there are really no points of intersection for the two graphs in the second quadrant.

Reset the Range to the default viewing rectangle by keying `Range` `SHIFT` `Mcl` `Range`. Now key `Prog` `0` `EXE` to redraw the functions. Using the Trace feature again, move the blinking cursor to the intersection in the first quadrant (see Figure 5.20).

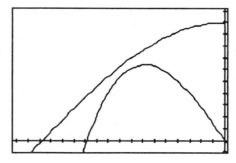

**Figure 5.19.**  Zoom-in by a factor of 5.

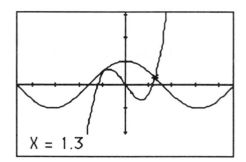

**Figure 5.20.**  Blinking cursor in the first quadrant.

Repeat the keying sequence (SHIFT) `Factor` `5` `:` `Prog` `0` `EXE` to zoom-in on this intersection by a factor of 5. To zoom-in more for greater accuracy, use the Trace feature to place the blinking cursor on the intersection, and then key `EXE` again to repeat the Factor command and draw the graphs in the new viewing rectangle. This process may be repeated up to the limits of machine accuracy. Figures 5.21 and 5.22 show the solution to the system of equations with error of at most 0.00001 .

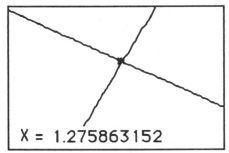

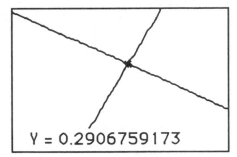

**Figure 5.21.** X-coordinate of the solution.    **Figure 5.22.**   Y-coordinate of the solution.

**5.3.5  The Plot and Line Features:** The Plot feature allows you to plot a point as a pixel on the graphics screen. For example, select the default viewing rectangle, $[-4.7, 4.7]$ by $[-3.1, 3.1]$, and then key in (SHIFT) $\boxed{\text{Plot}}$ $\boxed{.3}$ $\boxed{\text{SHIFT}}$ $\boxed{,}$ $\boxed{.5}$ $\boxed{\text{EXE}}$. Notice that the $x$-coordinate of the plotted point (blinking pixel) is displayed at the lower left-hand corner of the screen. Use $\boxed{\text{SHIFT}}$ $\boxed{X \leftrightarrow Y}$ to toggle between an $x$- and a $y$-coordinate readout. A plotted point can be moved using the cursor keys. Now try the following command: **Plot 1.75, 1.74** . The pixel selected is the one with coordinates nearest those of the input, $(1.8, 1.7)$ if you are still using the default viewing rectangle. Coordinate values falling outside the viewing rectangle will not be plotted. For instance, try plotting $(6, -3)$.

The **Line** command (SHIFT) $\boxed{\text{Line}}$ $\boxed{\text{EXE}}$ connects the two most recently plotted points. Try drawing a polygon with coordinate points as vertices using the Plot and Line commands.

The $\blacktriangle$ command may be used with the Graph, Plot, or Line commands to keep the graphics screen visible. This is illustrated in the next section.

**5.3.6  Zoom-In Using Interactive Viewing Rectangles:** One of the best features of the *Master Grapher* program for the microcomputer is its ability to zoom-in on a user-chosen rectangular subset of a viewing rectangle. This allows the user to see a picture of the actual area to be enlarged at each step of the zoom-in process.

The program presented in this section allows the Casio to do zoom-in by the same viewing-rectangle-within-viewing-rectangle method as *Master Grapher*. (The original concept of the Casio Zoom-In Program is due to Alan Stickney of Wittenberg University. The version presented here was written by Charles Vonder Embse of Central Michigan University.) You magnify an area containing a point of interest by drawing a rectangle around the point on the graphics screen. You then see the exact area that will become your next full screen. Figure 5.26 shows how the Casio screen looks after the user has drawn a rectangle around an area to be enlarged.

This Zoom-In Program takes advantage of the technique of placing functions to be graphed in a program (see paragraph 4"Program Storage Locations" in Section 5.3.4). In this section, Program 0 is used to store the Graph commands for the function or functions to be graphed. To begin you can graph the functions from Program 0. Trace, Factor, and Range features are all available to help you choose an appropriate viewing rectangle. Once you have the viewing rectangle you want and you are ready to begin the zoom-in process, the same Program 0 can be used as a subroutine for the Zoom-In Program.

When using the Zoom-In Program on the Casio, there is a continuous readout of the X or Y coordinates of the blinking point on the screen. This gives you immediate information about the part of the plane being viewed, the size of each pixel-to-pixel step, and the approximate location of points of interest on the screen.

## Zoom-In Program

| Code | Comments |
| --- | --- |
| Cls | [clears the graphics screen] |
| Lbl 1 | [label as position 1 for Goto statement at end of program] |
| Prog 0 | [executes Program 0 to graph the function(s)] |
| Plot ◢ | [puts the cursor in the middle of the screen and pauses] |
| X → A : Y → D | [coordinates of selected point into variables A and D] |
| Plot ◢ | [second cursor; pauses until EXE is keyed] |
| X → B : Y → C | [coordinates of selected point into variables B and C] |
| B > A ⇒ Goto 2 | *[checks order; if B > A, then go to position 2 ] |
| A → T: B → A : T → B | [swaps A and B if B < A; A must be smaller] |
| Lbl 2 | [label as position 2; comes here if B > A] |
| D > C ⇒ Goto 3 | *[checks order again; if D > C, then go to position 3] |
| C → T : D → C : T → D | [swaps C and D if D < C; C must be smaller] |
| Lbl 3 | [label as position 3; comes here if D > C] |
| Plot A, D | [plots point (A,D), the upper left corner of the view rectangle] |
| Plot B, D : Line | [plots (B, D), upper right corner; line from (A, D) to (B, D)] |
| Plot B, C : Line | [plots (B, C), lower right corner; line from (B, D) to (B, C)] |
| Plot A, C : Line | [plots (A, C), the lower left corner; line from (B, C) to (A, C)] |
| Plot A, D : Line ◢ | [replots (A, D); finishes viewing rectangle; pauses] |
| Range A, B, 1, C, D, 1 | [sets new Range to A, B, 1, C, D, 1] |
| Goto 1 | [go to position 1 and start again; regraphs function(s) in the new Range; starts the process again] |

The logical flow of the program is to graph the function in the current viewing rectangle and then locate a subrectangle to be enlarged. Identification of the subrectangle for zoom-in is done by finding the left, right, bottom, and top edges of the subrectangle. These four values correspond to A, B, C, and D, respectively, in the program. Once these values are established, a new Range is set using these values. For simplicity the scale marks on the axes are set to 1. After the diagonal corners of the viewing rectangle are set, the program automatically checks the order so that the smaller horizontal (left) value is placed in A and the larger (right) value is placed in B. A similar check is done for the vertical direction. The result is that the user may choose either pair of diagonal corners in any order. For convenience, the program should be stored as Program 9.

Let's use the Zoom-In Program to solve a problem.

*Problem*   Let $x$ be the side length of the square that is to be cut out from each corner of a 30-inch by 40-inch piece of cardboard to form a box with no top.

---

* The arrow used with the "Goto 2" and "Goto 3" statements in the program is an implication, or if-then, arrow and is accessed by pressing [SHIFT] followed by [7].

(1)  If the finished box is to have a volume of 1200 cubic inches, write two equations that can be solved simultaneously to determine the size of the square.

(2)  Draw complete graphs of each equation in the same viewing rectangle.

(3)  How many simultaneous solutions are there to your system in Part 1? Which of these solutions are also solutions to the problem situation?

(4)  Use zoom-in to determine all values of $x$ that produce a box of volume 1200 cubic inches.

*Solution to Part 1*   An algebraic representation of the volume of the box as a function of the side length $x$ is $V(x) = x(40 - 2x)(30 - 2x)$ or $V(x) = 4x^3 - 140x^2 + 1200x$. Either of these forms of the function can be used in the Casio. One of the above equations together with $V(x) = 1200$ are the two equations that can be solved simultaneously to determine the size of the square.

*Solution to Part 2*   Put the two functions to be graphed in Program 0. If you already have a function in this location, replace it with the commands:

$$\text{Graph Y} = 4X x^y 3 - 140X^2 + 1200X$$
$$\text{Graph Y} = 1200$$

Return to the Run mode, and run Program 0 to graph the functions. Make adjustments in the Range and regraph until you get a complete graph. The viewing rectangle $[-5, 30]$ by $[-500, 3500]$ works well (see Figure 5.23).

*Solution to Part 3*   There are three simultaneous solutions to the system of equations; however, only the two in the range $0 \le x \le 15$ are in the domain of the problem situation. If the length of the square corner cut away is greater than 15 inches, then the 30-inch side of the sheet of cardboard is completely cut away.

*Solution to Part 4*   With Program 0 and the Range set as shown in the Solution to Part 2, run the Zoom-In Program (stored as Program 9) by keying $\boxed{\text{Prog}}$ $\boxed{9}$ $\boxed{\text{EXE}}$. The function will be graphed again from Program 0. When the graph is finished a flashing point will appear in the center of the screen with the $x$-coordinate of that point at the bottom of the screen (see Figure 5.24). To switch between $x$- and $y$-coordinate readouts, key $\boxed{\text{SHIFT}}$ $\boxed{X \leftrightarrow Y}$.

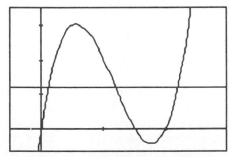

**Figure 5.23.** Graphs in [-5, 30] by [-500, 3500].

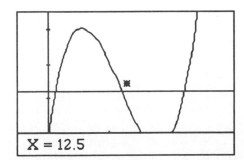

**Figure 5.24.** First point appears.

The flashing point in the middle of the screen is free to move to any point on the screen using the four cursor keys $\boxed{\Leftarrow}$, $\boxed{\Rightarrow}$, $\boxed{\Uparrow}$, and $\boxed{\Downarrow}$. This point will become one corner of your new viewing rectangle when [EXE] is pressed, and another flashing point will appear. This second point should be moved to the desired location of the diagonally opposite corner of the new viewing rectangle (see Figure 5.25). Once the second point is in position, press [EXE] to set the point and draw the viewing rectangle (see Figure 5.26).

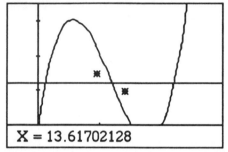

**Figure 5.25.** Second point in position.

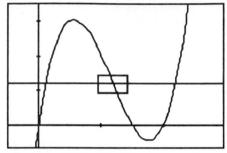

**Figure 5.26.** Viewing rectangle set.

When ready, press [EXE] again for the new viewing rectangle to be set and the graphs to be redrawn. After the new view of the graph is drawn, the flashing point again appears in the middle of the screen ready for the next zoom-in step (see Figure 5.27). At any time during the process of setting the corners of the viewing rectangle, the user can check the Range of the viewing screen by keying [Range]. The moving point can also be used to give the approximate location of the point of interest (an intersection in this case) by placing the point on the appropriate location on the screen (see Figure 5.28). This technique allows the user to make good estimates of the value being calculated and to keep in mind where the viewing rectangle is located in the plane. The change in the coordinates of the point as it moves across the screen gives a good indication of the size of the error in the estimate.

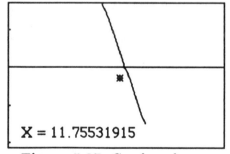

**Figure 5.27.** Graphs redrawn.

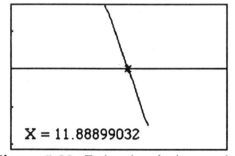

**Figure 5.28.** Estimating the intersection.

The Zoom-In Program can also make use of long, narrow viewing rectangles to stretch one dimension more than the other (see Figure 5.29). This technique makes intersections and local extreme points easier to read. Using a wide, short viewing rectangle as is seen in Figure 5.29 will tend to make the intersection appear more perpendicular than in the previous view. The vertical dimension will be stretched more than the horizontal. The size of the viewing rectangle is easy to control because of the point by point movement of the flashing pixel. Very small viewing rectangles can be selected; this speeds up the zoom-in process (see Figure 5.30).

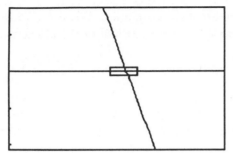

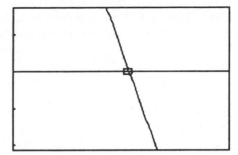

**Figure 5.29.** Long, narrow viewing rectangle.          **Figure 5.30.** Small viewing rectangle.

One solution that yields a box of 1200 cubic inches was shown in Figure 5.28. This solution of $x =$ 11.88899032 inches is accurate to .0005, despite the 9 decimal places shown on the Casio screen. Always check to see which digits change and which ones stay fixed as you move from point to point on the Casio graphics screen. This will allow you to estimate the error of a solution. To find the other solution to Part 4 of the problem, reset the viewing rectangle to $[-5, 30]$ by $[-500, 3500]$ and run Program 9 to return to the original complete graph shown in Figure 5.23. Then use the Zoom-In Program to approximate the solution. It should be about $x = 1.149$.

**5.3.7 Graphing Parametric Equations:** Parametric equations provide a powerful and flexible tool for representing curves and motion simulations in the Cartesian plane. Lines, including vertical lines, functions, inverses of functions, conic sections, and polar curves can all be represented by parametric equations. This is done by introducing a variable $t$, called a **parameter**, on which the values of $x$ and $y$ both depend. For motion simulations, the parameter $t$ can be thought of as representing time.

The following general Casio program can be used to graph any parametric curve. It is convenient to store this Parametric Curve Graphing Program as Program 8 (See paragraph 4 "Program Storage Locations" in Section 5.3.3). The Range is set outside the program; the parametric formulas are stored in Program 1, which acts as a subprogram; and the interval of $t$ values is set when the program is run.

**Parametric Curve Graphing Program**

```
"T MIN" ? → T
"T MAX" ? → B
"T INC" ? → H
Cls
Lbl 1
Prog 1
Plot X, Y
Line
T + H → T
T ≤ B ⇒ Goto 1
```

Pause to understand how the program works. Notice T starts at the T MIN that you specify and is incremented by H until T is equal to the T MAX value you specify. The X and Y values are computed within Program 1 using the current value of T; then, within the main program, the point (X, Y) is plotted and a line segment is drawn. (Occasionally, you may wish the plotted points **not** to be connected by line segments. In such cases, delete the Line command from you program.) The Casio flashes back and forth between the text window and the graphics window while the program is running. When the program has finished running, the Casio will show the text window, and the graph will be stored in the graphics window. The G $\leftrightarrow$ T command is used to view the completed graph.

Consider, for example, the parametric equations $x = t + 3$ and $y = t^2 - 2$ for $-3 \leq t \leq 3$. To graph this parametric curve, you would (a) store the statement T $+ 3 \rightarrow$ X : T$^2 - 2 \rightarrow$ Y as Program 1; (b) set the Range, say, to a viewing rectangle of $[-3, 9]$ by $[-3, 8]$; and (c) execute the Parametric Curve Graphing Program using $-3$ for T MIN, 3 for T MAX, and 0.05 for T INC. In this example, X takes on values in the interval $[0, 6]$, and Y in the interval $[-2, 7]$. The viewing rectangle of $[-3, 9]$ by $[-3, 8]$ was chosen with extra units in each direction (see Figure 5.31).

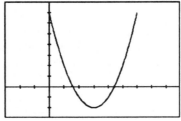

**Figure 5.31.** The parametric equations $x = t + 3$ and $y = t^2 - 2$ for $-3 \leq t \leq 3$, in the viewing rectangle $[-3, 9]$ by $[-3, 8]$.

**5.3.8   Graphing Polar Equations:**   Polar equations can be graphed using the Parametric Curve Graphing Program if you set up the subprogram called from the main program appropriately. In Section 5.3.7 the Parametric Curve Graphing Program was written so that Program 1 was the subprogram called from the main program.

*Problem*   Graph $r = 5 \sin 3\theta$.

*Solution   Make sure your Casio is in radian mode.* Then enter the following program as Program 1.

*Program 1*      **5sin 3T** $\rightarrow R$
                        R cos T $\rightarrow X$
                        R sin T $\rightarrow Y$

The portion in **boldface** type varies from one polar equation graphing problem to the next. Notice the variable T is used for the angle $\theta$.

To graph the polar curve, you would need to set the Range to an appropriate viewing rectangle, say, $[-7.5, 7.5]$ by $[-5, 5]$, and execute the Parametric Curve Graphing Program using 0 for T MIN, $\pi$ for T MAX, and 0.1 for T INC (see Figure 5.32).

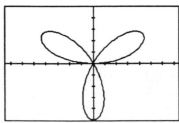

**Figure 5.32.** The rose $r = 5 \sin 3\theta$ in the viewing rectangle [-7.5, 7.5] by [-5, 5].

# Chapter 6

## Graphing with the Casio Power Graphic fx-7700G

The Casio Power Graphic fx-7700G calculator has many of the same features of the other members of the Casio graphing calculator line. Although their location on the keyboard has changed, the function of similar named keys remains basically the same. An example of this is the use of the $\boxed{\text{SHIFT}}$ and $\boxed{\text{ALPHA}}$ keys. Their use on the fx-7700G is the same as is described in Section 5.1. Rather than duplicating descriptions for the fx-7700G, similiar key functions have been referenced to the appropriate sections in Chapter 5, Calculating and Graphing with First Generation Casio Calculators. Also, the Owner's Manual has an excellent section for getting started with the fx-7700G.

### 6.1  Start Up

#### 6.1.1  Getting Turned On

**Power Up.** To turn the unit on, press $\boxed{\text{AC}}$ on. To turn the unit off, press $\boxed{\text{SHIFT}}$ $\boxed{\text{AC}}$ off. The power will automatically switch off after 6 minutes of inactivity. To restore functions, press $\boxed{\text{AC}}$ on.

**Adjust Contrast.** A system display will appear when the calculator is switched on. Press $\boxed{\text{MODE}}$. Press $\boxed{\triangleleft}$ to lighten the screen or $\boxed{\triangleright}$ to darken the screen.

#### 6.1.2  What You See—Display Windows: The fx-7700G displays both text and graphs on four different display windows.

(1)  The Text Display Window is the primary screen of the fx-7700G. On it you enter expressions and instructions and see results (except graphs).

**Figure 6.1.**  Text screen.

(2)  The Graphics Display Window contains 95 columns and 63 rows of pixels on which to display graphs. The graphics window can be scrolled (in all four directions) by pressing the appropriate cursor keys.

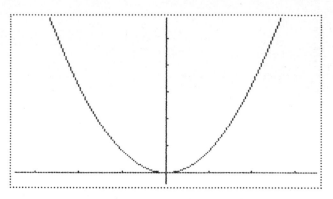

**Figure 6.2.**  Graphics screen.

(3)   The Range Setting Display Windows allow you to input the domain and range values for the graphics window.  Pressing $\boxed{\texttt{RANGE}}$ in either the text or the graphics window enters the range setting window and allows you to make adjustments to the viewing rectangle of the graph.

| | |
|---|---|
| minimum value of x − coordinate | *Range* |
| maximum value of x − coordinate | Xmin : −4.7 |
| distance between x − axis tick marks |   max : 4.7 |
| minimum value of y − coordinate |   scl : 1. |
| maximum value of y − coordinate | Ymin : −3.1 |
| distance between y − axis tickmarks |   max : 3.1 |
| |   scl : 1. |
| | INIT |

**Figure 6.3.**  Rectangular coordinate range screen.

Pressing $\boxed{\texttt{RANGE}}$ again brings up the second range screen.

| | |
|---|---|
| | Range |
| | T, $\theta$ |
| minimum value of T $\theta$ |   min : 0 |
| maximum value of t $\theta$ |   max : $4\pi$ |
| pitch of T $\theta$ | ptch : $\pi/50$ |
| | INIT |

**Figure 6.4.**  Polar coordinate range screen.

(4)   The Mode Display Windows appear on the screen when the fx-7700G is first turned on.  They specify the kind of functions the Casio is ready to perform.  Because the Casio can do so many things, it may be necessary to change one or more of the mode settings before you start a calculation.  To make any necessary changes, press the $\boxed{\texttt{MODE}}$ key to see the first set of options and then the $\boxed{\texttt{SHIFT}}$ key for additional options.  A summary of all the mode options is given below.  Press $\boxed{\texttt{MODE}}$ to display Mode Menu 1.

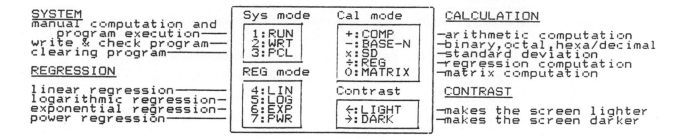

**Figure 6.5.** Mode Menu 1 screen.

Pressing $\boxed{\text{SHIFT}}$ will now display Mode Menu 2.

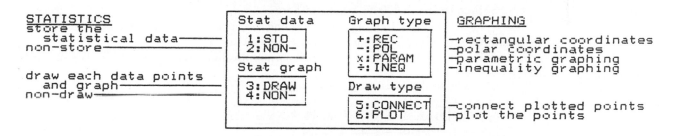

**Figure 6.6.** Mode Menu 2 screen.

You are encouraged to confirm which modes are active before calculating, graphing, or programming by pressing $\boxed{\boxed{\text{M}}\text{Disp}}$. The mode display window will remain on the screen while this key is depressed.

We will begin in the RUN mode of the system, COMPutation mode of the calculation, and the RECtangular coordinate graphing mode. To make these selections, press $\boxed{\text{MODE}}$ followed by $\boxed{1}$ for the RUN mode. Then press $\boxed{\text{MODE}}$ again followed by $\boxed{+}$ for the COMP mode. Finally, press $\boxed{\text{MODE}}$ $\boxed{\text{SHIFT}}$ followed by $\boxed{+}$ for REC to set the graph mode to rectangular coordinate graph type. Before beginning the next section, check your system display window to confirm:

```
     RUN   /   COMP
  G — type  :   REC/CON
  angle     :   Deg
  display   :   Nrm2
```

**Figure 6.7.** Mode display.

In each of the four display windows, the fx-7700G has several types of cursors. In most cases, the appearance of the cursor indicates what will happen when you press the next key. The most common cursors seen are given in the following table.

| Cursor | Appearance | Meaning |
|---|---|---|
| Entry cursor | Blinking underline | The next keystroke is entered at the cursor over-writing any character |
| Insert cursor | Open blinking rectangle | The next keystroke is inserted at the cursor. |
| SHIFT cursor | Highlighted blinking S | The next keystroke is a second function. |
| ALPHA cursor | Highlighted blinking A | The next keystroke is an alpha character. |
| Graphics cursor | Blinking plus sign | Designates a point on the graphics display window. |

**Table 6.1**

## 6.2  Getting Going (Pushing Buttons)

**6.2.1  Calculations:** In general the keystroke sequence necessary to evaluate simple expressions on the fx-7700G is the same as for the other members of the Casio line. Section 5.1.4 describes the keystroke sequence. A few features unique to the fx-7700G are listed here.

The subtraction key can be used as the negation function key if it is not the first operation in an expression. This is especially useful when entering negative range values.

The fraction notations key $\boxed{\text{a b/c}}$ allows entry of fractions and mixed numbers. The $\boxed{\text{d/c}}$ key converts to an improper fraction.

*Example*  $3/4 + 1/3$

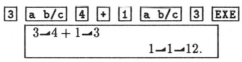

$$\boxed{3}\ \boxed{\text{a b/c}}\ \boxed{4}\ \boxed{+}\ \boxed{1}\ \boxed{\text{a b/c}}\ \boxed{3}\ \boxed{\text{EXE}}$$

|  |
|---|
| $3\llcorner 4 + 1\llcorner 3$ |
| $1\llcorner 1\llcorner 12.$ |

**Figure 6.8.**  Adding fractions.

The symbol "$\llcorner$" separates the parts of the fraction. So $1\llcorner 1\llcorner 12$ is the same as $1\frac{1}{12}$.
Press $\boxed{\text{a b/c}}$ to convert to a decimal.

|  |
|---|
| $3\llcorner 4 + 1\llcorner 3$ |
| $1.083333333$ |

**Figure 6.9.**  Convert fraction to decimals.

Press $\boxed{\text{SHIFT}}\ \boxed{\text{a b/c}}$ to convert to an improper fraction.

|  |
|---|
| $3\llcorner 4 + 1\llcorner 3$ |
| $13\llcorner 12.$ |

**Figure 6.10.**  Convert to improper fractions.

A distinctive feature of the fx-7700G is that many of the operations are not on the keyboard but instead are stored within function menus, shown in green above certain keys throughout the keyboard. To access any of the menus, press $\boxed{\text{SHIFT}}$ followed by the desired function key. A group of up to six functions will then be displayed across the bottom of the screen. To select a choice, press the corresponding $\boxed{\text{F}}$ key. For example, to compute the absolute value of 2 minus five factorial requires the use of operations stored in the function menus. For $|2| - 5!$ enter $\boxed{\text{SHIFT}}\ \boxed{\text{MATH}}$ to access the math menu. Now enter $\boxed{\text{F3}}$ to access the numeric menu, and then $\boxed{\text{F1}}$ for the absolute value function. The rest of the key sequence follows. $\boxed{2}\ \boxed{-}$ $\boxed{5}\ \boxed{\text{SHIFT}}\ \boxed{\text{MATH}}\ \boxed{\text{F2}}\ \boxed{\text{F1}}\ \boxed{\text{EXE}}$.

$$\boxed{\begin{array}{l} \text{Abs } 2 - 5! \\ \hfill -118 \end{array}}$$

**Figure 6.11.** Using menus.

Notice that the factorial function is also stored in the math menu.

A diagram is given below to show the most commonly used menus in the computation mode. Use it as a reference to know what operations each menu key accesses. A more detailed listing of menus is included in the Owner's Manual.

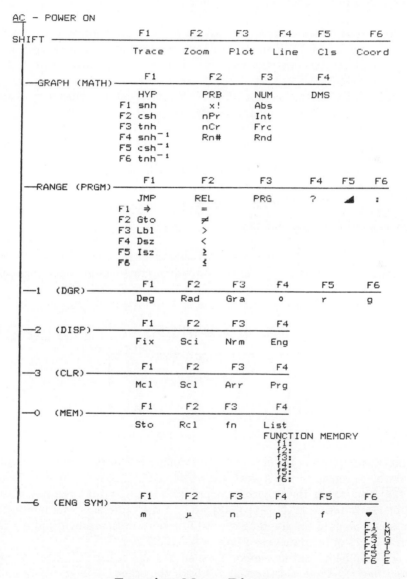

**Function Menu Diagram**

When you enter a function menu a group of up to six functions will be displayed across the bottom of the screen. To select a choice, press the corresponding $\boxed{F}$ key. To exit a menu, or backtrack when you are several layers into a menu, press $\boxed{\text{Pre}}$. The next two examples use function menus. To change from radian measure to degree measure you must use the DRG menu. Enter $\boxed{\text{SHIFT}}$ $\boxed{\text{DRG}}$. The choices for this menu are now listed at the bottom of the screen. Enter $\boxed{\text{F1}}$ to select radians and $\boxed{\text{EXE}}$ to enter it in the mode setting. Holding down the $\boxed{\text{M Disp}}$ key verifies the change in angle measurement. To compute hyperbolic sine, access the math menu and then the hyperbolic function. To compute sinh(.387) enter $\boxed{\text{SHIFT}}$ $\boxed{\text{MATH}}$ $\boxed{\text{F1}}$ $\boxed{\text{F1}}$ $\boxed{.387}$ $\boxed{\text{EXE}}$. You find it equal to .3967326982.

**6.2.2 Special Functions:** When editing expressions you can incorporate the cursor keys (as REPLAY keys), and the INSERT and DELETE keys. These keys are very similiar in operation to the other Casio models. Section 5.1.3 explains the operations as well as how to deal with error messages for the Casio graphing calculators.

You can obtain sequential execution of a number of individual statements by using multistatements. There are three different ways to form multistatements.

(1)  The colon key $\boxed{:}$ accessed by $\boxed{\text{SHIFT}}$ $\boxed{\text{PRGM}}$ $\boxed{\text{F6}}$ connects statements left to right without stopping.

(2)  The newline key $\boxed{\leftarrow}$ accessed by $\boxed{\text{SHIFT}}$ $\boxed{\leftarrow}$ operates like the colon when placed at the end of a line. It moves the cursor to the next line but connects the statements like a colon.

(3)  The display key $\boxed{\blacktriangleleft}$ accessed by $\boxed{\text{SHIFT}}$ $\boxed{\text{PRGM}}$ $\boxed{\text{F5}}$ stops execution and displays results to that point. You can resume execution by pressing $\boxed{\text{EXE}}$.

By entering Statement$\boxed{:}$Statement or Statement$\boxed{\leftarrow}$Statement or Statement$\boxed{\blacktriangleleft}$Statement, multistatements are formed allowing sequential execution of individual statements. This particular strategy is very helpful in evaluating functions.

For example, to evaluate $A^2 + 2A + 3$ when A is 2, 3, or -1, enter

$\boxed{\text{SHIFT}}$ $\boxed{\text{PRGM}}$ $\boxed{\text{F4}}$ $\boxed{\rightarrow}$ $\boxed{\text{ALPHA}}$ $\boxed{\text{A}}$ $\boxed{\text{F6}}$ $\boxed{\text{ALPHA}}$ $\boxed{\text{A}}$ $\boxed{\text{SHIFT}}$ $\boxed{x^2}$ $\boxed{+}$ $\boxed{2}$ $\boxed{\text{ALPHA}}$ $\boxed{\text{A}}$ $\boxed{+}$ $\boxed{3}$ $\boxed{\text{EXE}}$.

$$
\begin{array}{l}
? \rightarrow A : A^2 + 2A + 3 \\
? \\
1 \\
\phantom{xxxxxxxxxxxxxxx} 6
\end{array}
$$

**Figure 6.12.** Evaluating functions.

The question mark is requesting the first value to evaluate in the expression. By entering $\boxed{1}$ $\boxed{\text{EXE}}$ you are telling the calculator to replace every A with the value of 1. Of course, the expression simplifies to 6.

If you enter $\boxed{\text{EXE}}$ you invoke the replacement program again and it responds with ?. Enter $\boxed{2}$ $\boxed{\text{EXE}}$ to obtain the value of 11. Entering $\boxed{\text{EXE}}$ $\boxed{-1}$ $\boxed{\text{EXE}}$ we find the expression evaluated at -1 to be 2.

Multistatements can be used for graphing several functions in the same graphing window.

There are several types of memories used on the fx-7700G.

Numerical values can be stored in memory, called storage registers. The storage registers for the fx-7700G can be described the same as in Section 5.1.5 for the other Casio graphing models.

Functions can be stored in a function memory on the fx-7700G. You can store up to six functions in memory for instant recall when you need them. To access the function memory press $\boxed{\text{SHIFT}}$ $\boxed{\text{F}}\boxed{\text{MEM}}$.

**Figure 6.13.** Function memory.

Press the appropriate $\boxed{\text{F}}$ key to store a function, recall a function, display a list of the stored functions, or specify the input as a function when using in an operation. For example, store $x^3 + 2x^2$ in the first function memory and then graph $y = x^3 + 2x^2$. Begin by entering $\boxed{\text{x},\theta,\text{T}}$ $\boxed{x^y}$ $\boxed{3}$ $\boxed{+}$ $\boxed{2}$ $\boxed{\text{x},\theta,\text{T}}$ $\boxed{\text{SHIFT}}$ $\boxed{x^2}$ $\boxed{\text{SHIFT}}$ $\boxed{\text{F}}\boxed{\text{MEM}}$ $\boxed{\text{F1}}$ $\boxed{1}$. Use the default range settings. The function $y = x^3 + 2x^2$ has been stored in function memory #1. Now to graph $y = x^3 + 2x^2$ key in $\boxed{\text{GRAPH}}$ $\boxed{\text{F3}}$ $\boxed{1}$ $\boxed{\text{EXE}}$.

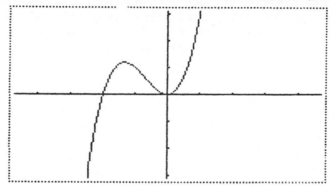

**Figure 6.14.** $y = x^3 + 2x^2$ [-4.7, 4.7] by [-3.1, 3.1].

This brings us to the graphing capabilities of the fx-7700G.

## 6.3 Graphing with the FX-7700G

**6.3.1 The Set-Up:** Special keys on the fx-7700G are used to graph functions. When you press $\boxed{\text{GRAPH}}$, the text screen is displayed with Graph Y= and awaits the entry of a function to graph. When you press $\boxed{\text{G}\leftrightarrow\text{T}}$, the display switches from the graphics display window (shows the coordinate grid) to the text display window (shows graph formulas) or vice versa. When you press $\boxed{\text{RANGE}}$, an edit screen is displayed where you define the viewing rectangle for the graph. In addition to these keys, the $\boxed{\text{F}}$ keys at the bottom of the screen access six functions (Trace, Zoom, Plot, Line, Cls, Coord) used with graphing. When you press $\boxed{\text{F1}}$ (Trace), you can move the cursor along a graphed function and display the X and Y coordinate values of the cursor location on the function. When you press $\boxed{\text{F2}}$ (Zoom), you access a menu of instructions at the bottom of the screen that allows you to change the viewing rectangle. Pressing $\boxed{\text{F3}}$ (Plot) allows you to plot a point anywhere on a graph. Pressing $\boxed{\text{F4}}$ (Line), allows you to link two points with a straight line. Pressing $\boxed{\text{F5}}$ (Cls), allows you to clear the graphics screen. Pressing $\boxed{\text{F6}}$ (Coord) changes the coordinate display from both X and Y to X coordinate only then to Y coordinate only and then back to both X and Y coordinates.

Because the fx-7700G is capable of so many different types of graphing it is crucial that you set the mode to correspond with the type of graphing you are currently using. For this section we will use the rectangular coordinate system. In addition, we must choose the way the graph will be drawn—only plotted points or the points connected. To select the type of graph to be drawn, access the second mode window by entering MODE SHIFT and make a selection—5 to connect, and 6 to plot only—followed by EXE. Check your system display window to confirm that this is your mode.

```
   RUN   /   COMP
 G — type  :  REC/CON
   angle   :  Deg
 display   :  Nrml
```

**Figure 6.15.** Mode display.

Now check the range values to determine the size of the viewing rectangle. The Range key allows you to choose the viewing rectangle that defines the portion of the coordinate plane that appears in the display. The values of the RANGE variable determine the size of the viewing rectangle and the scale units for each axis. You can view and change the values of the RANGE variable almost any time. Press RANGE to display the RANGE variables edit screen. The values shown here on the RANGE edit screen are the standard default values in radian mode.

```
Rectangular Range Screen        Polar Coordinate Range Screen

  Range                            Range
  Xmin : −4.7                      T, θ
    max : 4.7                        min : 0
    scl : 1                          max : 4π
  Ymin : −3.1                      ptch : π/50
    max : 3.1
    scl : 1.                       INIT
  INIT
```

**Figure 6.16.** Default values for both range screens.

Note that the parameters can be entered using the $\pi$ symbol, but the calculator converts them to decimal equivalences.

To obtain this default range setting quickly, enter RANGE F1. This operation initializes the range settings. The second range screen is used for polar and parametric equation parameters. To escape the range setting window, you must press RANGE two times. The first entry RANGE will take you to the polar coordinate/parametric range screen. Pressing RANGE again takes you back to the last displayed screen, text or graph.

To change one of the RANGE values, move the cursor to the desired line and type in the new value. You may also edit an existing value to produce a new value.

Examine the coordinate grid now on the graphic viewing screen and verify that its axes represent the standard viewing rectangle of [-4.7, 4.7] scl 1 by [-3.1, 3.1] scl 1.

### 6.3.2  Generating Graphs:

**Graphing Built-in Functions.**  The Casio calculator provides built-in graphs for 40 basic functions. To automatically generate the graph of any one of these functions press the GRAPH key together with the desired function key. Built-in functions have preselected ranges set automatically by the calculator. These ranges ensure that a complete and representative graph picture is produced.

To obtain the graph of the sine function simply press GRAPH sin EXE . (Note: Do not enter the x.) Check the range values by pressing the RANGE key. The range should be the decimal equivalent for $[-2\pi,$ $2\pi]$ scl $\pi$ by [-1.6, 1.6] scl .5.

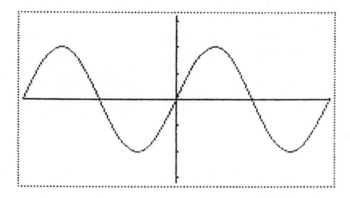

**Figure 6.17.**  $y = \sin x$ $[-2\pi, 2\pi]$ by [-1.6, 1.6].

Press the RANGE key twice to get back to the graph, then key in G ↔ T to return to the text window. Now key in GRAPH SHIFT ∛ EXE to see the built-in cube root function. What are the Range values now? Have they changed? They should have. They should now be essentially [-9, 9] scl 2 by [-2.5, 2.5] scl 1.

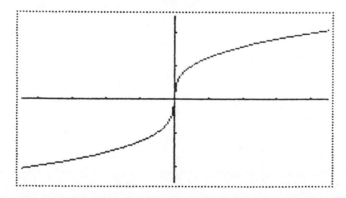

**Figure 6.18.**  $y = \sqrt[3]{x}$ [-9, 9] by [-2.5, 2.5].

**Clearing the Graphics Window.**  Whenever any of the Range settings are changed, the graphics display window is automatically cleared.  At times you may wish to clear all graphs from the graphics window without changing the viewing rectangle or scale marks. To do this, press SHIFT Cls EXE . You can check to see that the graphics window has been cleared by pressing $\boxed{\text{G} \leftrightarrow \text{T}}$. The graphs are removed, but the scaled axes are not. It is not necessary to clear the graph of one built-in function before the graph of a second built-in function is executed. The previous graph is automatically cleared from the coordinate grid before the new function is drawn—you can only graph one built-in function at a time.

Unless you want one or more graphs to be drawn on the same coordinate grid, the above clearing sequence must be entered or the range settings changed before each new graph is executed.

**Scrolling.** You can scroll the entire graph/axes in four directions by pressing any of the cursor keys. Scrolling allows you to see different sections of your graph not necessarily visible on the initial graph's screen. You can use scroll with built-in and user-generated graphs but not with polar coordinate or parametric graphs. You cannot scroll when TRACE is active.

Scroll the built-in natural logarithm function in all directions. Press $\boxed{\text{GRAPH}}$ $\boxed{\text{ln}}$ $\boxed{\text{EXE}}$ for the graph. Then press any of the cursor keys to scroll the screen.

**Graphing User-Generated Functions.** The secret to generating the graph of a given formula within a particular viewing rectangle is to press the $\boxed{\text{GRAPH}}$ key and then enter the function formula in terms of X, $\theta$, or T. (The $\boxed{\text{X},\theta,\text{T}}$ key lets you enter the variable quickly without pressing $\boxed{\text{ALPHA}}$.) Unlike built-in graphs, user-generated graphs do not clear automatically as a new graph is executed.

It is also important to remember that whenever you execute a user-generated graph, if you do not reset the range values, the size of the viewing rectangle will be determined by the last set of range settings entered. Although a given viewing rectangle may produce a great picture of one function, it may be very inadequate for the next function you wish to graph.

As an example of this, look at the function $y = 2x^4 + 3x^3 - 4x^2 + 5x - 6$. with a range of [-30, 30] scl 3 by [-30, 30] scl 3. To enter the function

Press:      $\boxed{\text{GRAPH}}$ $\boxed{2}$ $\boxed{\text{X},\theta,\text{T}}$ $\boxed{x^y}$ $\boxed{4}$ $\boxed{+}$ $\boxed{3}$ $\boxed{\text{X},\theta,\text{T}}$ $\boxed{x^y}$ $\boxed{3}$ $\boxed{-}$ $\boxed{4}$ $\boxed{\text{X},\theta,\text{T}}$ $\boxed{\text{SHIFT}}$ $\boxed{x^2}$ $\boxed{+}$ $\boxed{5}$ $\boxed{\text{X},\theta,\text{T}}$ $\boxed{-}$ $\boxed{6}$ $\boxed{\text{EXE}}$.

**Figure 6.19.**   $y = 2x^4 + 3x^3 - 4x^2 + 5x - 6$ [-30, 30] by [-30, 30].

The shape of the graph is hard to visualize since the graph seems to be compressed. Press the $\boxed{\text{RANGE}}$ key again and reset the range values to [-5, 5] scl 1 by [-25, 300] scl 10. After entering the last scale value, press $\boxed{\text{EXE}}$ $\boxed{\text{RANGE}}$ $\boxed{\text{EXE}}$ and the graph will be automatically redrawn in the new viewing rectangle.

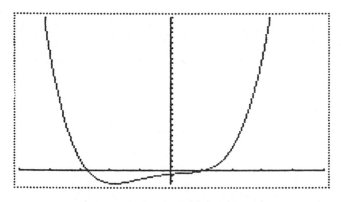

**Figure 6.20.**  [-5, 5] by [-25, 300].

The second viewing rectangle clearly produced a better, easier to interpret graphic display for this particular function. Because the Casio is limited in both display size and screen resolution, it is generally a good idea to graph a function using several different viewing rectangles to get both the "big picture" as well as the fine details. Clear the graphics display window using the $\boxed{\texttt{Cls}}$ key and graph $f(x) = x^3 + 7x^2 + 9x$ over the following different range settings.

$$[-30, 30] \text{ scl } 3 \text{ by } [-100, 100] \text{ scl } 20$$
$$[-10, 10] \text{ scl } 1 \text{ by } [-10, 30] \text{ scl } 3$$
$$[-6, 4] \text{ scl } .5 \text{ by } [-4, 16] \text{ scl } 1$$

Compare the graphs.

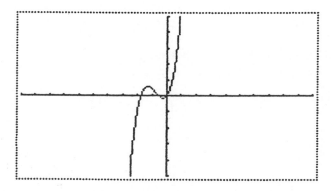

**Figure 6.21.**  $y = x^3 + 7x^2 + 9x$
[-30, 30] by [-100, 100].

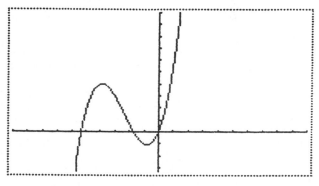

**Figure 6.22.**  [-10, 10] by [-10, 30].

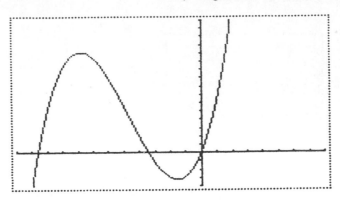

**Figure 6.23.** [-6, 4] by [-4, 16].

Determining an appropriate viewing rectangle will always involve some amount of trial and error however, there are several ways in which you may be able to reduce the amount of trial and minimize the amount of error·

(1)  Use the range settings developed for the 40 built-in functions as a guide when graphing similar functions.

(2)  Since the viewing screen is a rectangle, try a ratio of x/y = 3/2 when setting the range if you're plotting shapes that must look a certain way (i.e., circles, perpendicular lines). The standard viewing rectangle [-4.7, 4.7] scl 1 by [-3.1, 3.1] scl 1 which can be accessed by pressing [F1] when the range setting window is displayed, will give you this ratio.

(3)  One author suggests using the order of the polynomial to determine an appropriate ratio of x/y. For example, set the ratio of x/y = a/a² for a quadratic and for a cubic let the ratio of x/y = a/a³.

**6.3.3 Overlaying, Tracing, and Zooming:** Now that you are familiar with how to set Range values, the next steps in graphing on the Casio are to overlay graphs and to use the Trace and Zoom features.

**Overlaying graphs.** The multistatement keys (described in Section 6.4.2) are used to graph several functions—one after the other in the same viewing rectangle. This example shows two functions graphed together.

Clear the graphics display window, [SHIFT] [F5] [EXE], enter the following range values: x[-10, 10] scl 2 by y[-20, 20] scl 2 and then graph each of the following functions $f(x) = x^2$ and $g(x) = 2x + 8$ using the keystroke sequence outlined below:

Press:      [GRAPH] [X,$\theta$,T] [SHIFT] [$X^2$] [SHIFT] [↵] [GRAPH] [2] [X,$\theta$,T] [+] [8] [EXE]

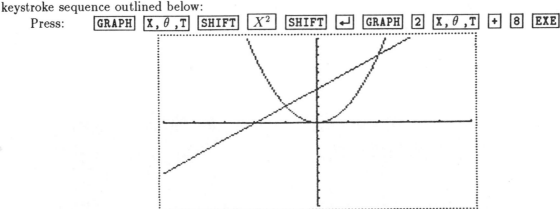

**Figure 6.24.** $y = x^2$ and $y = 2x + 8$ overlaid.

Now reset the range settings to [-5, 5] scl 1 by [-20, 20] scl 2 and press $\boxed{\texttt{EXE}}$ $\boxed{\texttt{RANGE}}$ $\boxed{\texttt{EXE}}$. Both graphs are now redrawn within the new viewing rectangle.

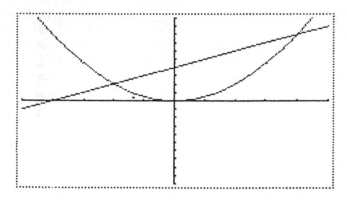

**Figure 6.25.** $y = x^2$, $y = 2x + 8$ [-5, 5] by [-20, 20].

**Trace.** There will be times when you will want to know the coordinates of a given point on a graph. Pressing the $\boxed{\texttt{F1}}$ key while in the graphics window will allow you to move a blinking pointer along the last graph executed and identify the x and y value of each point on that graph. Enter the following range values: [-8, 8] scl 1 by [-6, 6] scl 1 and then graph $f(x) = x + 2$.

Press:     $\boxed{\texttt{GRAPH}}$ $\boxed{\texttt{X},\theta,\texttt{T}}$ $\boxed{\texttt{+}}$ $\boxed{\texttt{2}}$ $\boxed{\texttt{EXE}}$
Press:     $\boxed{\texttt{F1}}$.

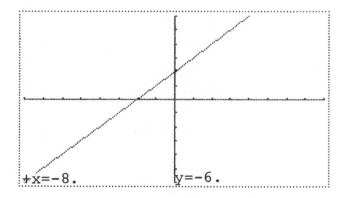

**Figure 6.26.** Using the trace feature.

If you look carefully, you will notice that the bottom left-most dot of the graph is blinking. The coordinates of this blinking dot correspond to the coordinates displayed on the screen. Pressing the left and right cursor keys will move the blinking pointer along the graph and simultaneously change the values displayed on the screen. (Note: the cursor moves from dot to dot on the screen. When you move the cursor to a dot that appears to be "on" the function, it may be near, but not on, the function. The coordinate value is accurate to within the width of the dot (see Section 2.4).)

For overdrawn graphs using multistatements, the traced pointer can be moved between the graphs. Graph and find the point(s) of intersection for $y = x^2$ and $y = .5x + 1$. Using the default viewing rectangle

Press: $\boxed{\text{GRAPH}}$ $\boxed{\text{X},\theta\,,\text{T}}$ $\boxed{\text{SHIFT}}$ $\boxed{x^2}$ $\boxed{\text{SHIFT}}$ $\boxed{\hookleftarrow}$ $\boxed{\text{GRAPH}}$ $\boxed{.5}$ $\boxed{\text{X},\theta\,,\text{T}}$ $\boxed{+}$ $\boxed{1}$ $\boxed{\text{EXE}}$.

Now press $\boxed{\text{F1}}$ to Trace. The pointer appears on the graph drawn by the last function in the multistatement. The tracer can be moved between graphs entered as multistatements by pressing the up or down cursor keys. Find the two points of intersection for these graphs.

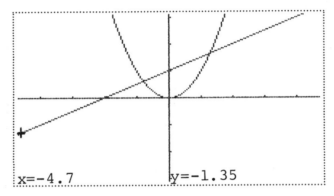

**Figure 6.27.** Intersection of $y = x^2$ and $y = .5x + 1$.

**Zoom.** This function allows you to enlarge or reduce portions of the graph or the entire graph itself. While in the graphic display window press $\boxed{\text{F2}}$ to access the Zoom Menu. The Zoom Menu options are listed at the bottom of the text screen.

$\boxed{\text{F1}}$ The Box Function lets you cut out a specific section of a graph for zooming. Graph $y = 2\cos x$ and $y = -x^3 + 2x$ on the default viewing rectangle using a multistatement command.

Press: $\boxed{\text{GRAPH}}$ $\boxed{2}$ $\boxed{\cos}$ $\boxed{\text{X},\theta\,,\text{T}}$ $\boxed{\text{SHIFT}}$ $\boxed{\hookleftarrow}$ $\boxed{\text{GRAPH}}$ $\boxed{-}$ $\boxed{\text{X},\theta\,,\text{T}}$ $\boxed{x^y}$ $\boxed{3}$ $\boxed{+}$ $\boxed{2}$ $\boxed{\text{X},\theta\,,\text{T}}$ $\boxed{\text{EXE}}$

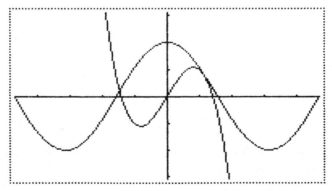

**Figure 6.28.** $y = 2\cos x$, $y = -x^3 + 2x$ [-4.7, 4.7] by [-3.1, 3.1].

Now press $\boxed{\text{F2}}$ to access the Zoom menu followed by $\boxed{\text{F1}}$ for the Box function.

This lets you draw a box anywhere on the screen in which to magnify the graph for a new viewing. The cursor is in the middle of the screen. Use the cursor arrow keys to move the cursor from the middle of the graph to where you *want one corner* of the new viewing rectangle to be. Press $\boxed{\text{EXE}}$. Next, move the cursor to the diagonally opposite corner of the desired viewing rectangle. The outline of the new viewing rectangle is drawn as you move the cursor. (See Figure 6.29.) Press $\boxed{\text{EXE}}$ to accept the cursor location as the second corner of the box. The graph is replotted immediately using the box outline as the new viewing rectangle. Use as many zoom-ins as necessary to obtain the desired picture on the graph screen, Figure 6.30.

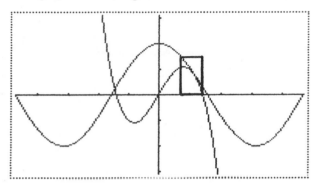

**Figure 6.29**                    **Figure 6.30**

To return to the original graph press $\boxed{\text{F2}}$ for the Zoom menu and $\boxed{\text{F5}}$ for the ORG key. This allows a quick and easy way to return to the original graph after zooming or scrolling.

$\boxed{\text{F2}}$ The FCT factor function determines the scale of the magnification for the Zoom In or Zoom Out features. Before using Zoom In or Zoom Out, you can review or change the zoom factors. Zoom factors are positive numbers (not necessarily integers) greater than or equal to 1. If a factor value between zero and one is chosen, the viewing rectangle will be increased to reduce or "zoom out" on the graph. Choosing a factor value greater than one will produce a smaller viewing rectangle to magnify or "zoom in" on a particular portion of the graph. The same factor, or two separate factors, can be specified to magnify or reduce each axis.

To review the current values of the zoom factors, select $\boxed{\text{F2}}$ (FCT) from the ZOOM menu. The Zoom Factors screen appears. The default factor is 2 for each direction. If the factors are not what you want, change them.

$\boxed{\text{F3}}$
and
$\boxed{\text{F4}}$ The xf (Zoom-In) and $x^1/f$ (Zoom-Out) keys are the built-in zooming keys. The xf key $\boxed{\text{F3}}$ causes the graph to be enlarged about the center of the screen $\Rightarrow$ Zooming-in. The $x^1/f$ key $\boxed{\text{F4}}$ causes the graph to be reduced about the center of the screen $\Rightarrow$ Zooming-out. When incorporating the Trace function with the Zoom menu the zooming will be done with respect to the traced point. With the functions of $y = 2\cos x$ and $y = -x^3 + 2x$ graphed over the default range setting press the xf function key $\boxed{\text{F3}}$. (Be sure the Zoom Factors $\boxed{\text{F2}}$ are set at the default of 2 in both directions.) What happens to the graph? What happens if you press the $x^1/f$ function key $\boxed{\text{F4}}$?

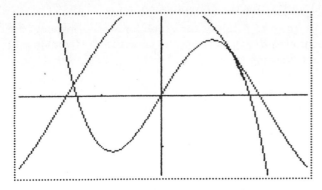

**Figure 6.31.** Using xf Zoom-In. 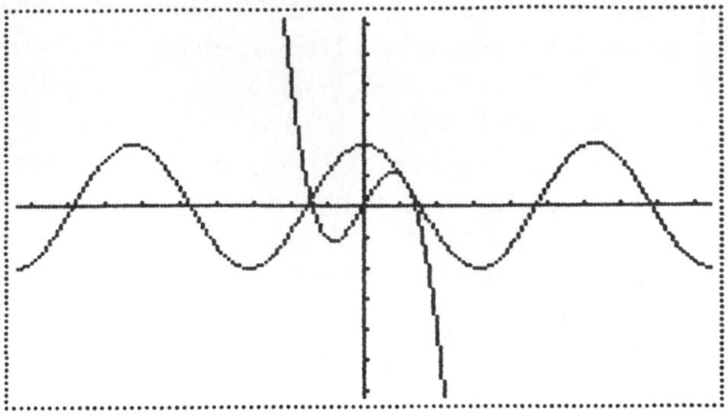      **Figure 6.32.** Using x$^1$/f Zoom-Out.

[F5] The ORG function returns the graph to its original settings

Using the Casio features just described, you can graphically solve equations, inequalities, and systems of equations. These features open the door to solving extreme-value (max/min) problems and to determining intervals over which a function is increasing or decreasing. The same methods apply regardless of the functions involved. The equation $\cos x = \tan x$ can be solved by the same graphical method that would be used to solve $2x = 6$.

*Problem*　Solve $\cos x = \tan x$ for $0 \le x \le 1$.

*Note*　This same problem is worked out in Section 5.2.2 for the other Casio models.

*Solution*　Clear the graphics screen and reset the default viewing rectangle by keying in [RANGE] [F1] [RANGE] [RANGE]. Then enter the function by pressing [GRAPH] [cos] [X,θ,T] [SHIFT] [↵] [GRAPH] [tan] [X,θ,T]. The text screen should look like this:

$$\boxed{\begin{array}{l} \texttt{Graph } Y = \cos X \\ \texttt{Graph } Y = \tan X \end{array}}$$

**Figure 6.33**

Edit if necessary and then press [EXE]. Watch carefully. Notice that the cosine function is drawn first, and then the tangent function is overlaid.

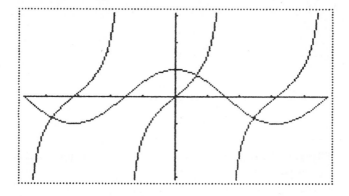

**Figure 6.34.** $y = \cos x$ and $y = \tan x$ overlaid.

You can approximate the coordinates of any point on either graphed function with the TRACE feature. Press [TRACE]. Locate the blinking pixel at the left most point of the last graphed function. Use the cursor keys repeatedly to move the blinking pixel to the point of intersection of the graphs of $Y = \cos X$ and $Y = \tan X$ that lies between $x = 0$ and $x = 1$. The ordered pair should be (0.7, 0.7648421).

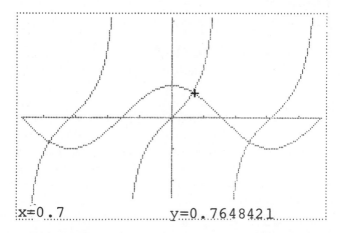

x=0.7                              y=0.7648421

Now press [F2] (zoom) and then [F3] (xf). See Figure 6.36. Notice that the fx-7700G uses the traced-to point as the center of the new, smaller, viewing rectangle. The automatic zoom factor is 2 in both the x and y directions. Press [F2] [F3] again. See Figure 6.37. If you did not use Trace, the Casio zooms in about the center of the current viewing rectangle. Without using Trace and without comparing to any graph given here, zoom in three more times (for a total of 5 zoom-in steps). The final graph should be like Figure 6.38.

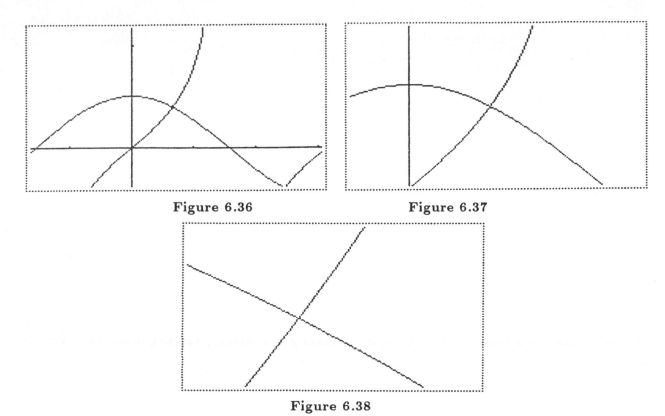

Figure 6.36                              Figure 6.37

Figure 6.38

Now use Trace again to locate the new, improved approximation for the point of intersection. To zoom-out automatically press [F2] [F4].

For a different approach to the same problem use the Box Function to find the point of intersection. Return to the original settings by pressing [F2] and then [F5]. To access the Box Function, press [F2] and then [F1]. Move the blinking cursor to the point you wish to set as the top left corner of the box and [EXE] to set the point. Now move the blinking cursor right and down with the cursor keys to set the bottom right corner of the box, followed by [EXE]. When the box is set, the graph will be automatically redrawn within a viewing rectangle defined by the parameters of the box just designed. Trace for the point of intersection.

When you first begin using the fx-7700G, you may be more comfortable zooming-in and zooming-out by changing the Range settings by hand. Eventually you will probably use a variety of methods depending on the situation.

## 6.4  Examples of Different Kinds of Graphing

**6.4.1  Stored Functions:** Begin by storing the functions in memory, using the method described in Section 6.2.5. Store $x^3 - 1$ and $x - x^2$ into the function memory. Set the range at [-4, 4] scl 1 [-10, 10] scl 1. To graph, you must recall the [F]MEM, enter [F3] to call the function and then the number it is stored under.

Press:      [X,θ,T] [$x^y$] [3] [−] [1] [SHIFT] [F]MEM [F1] [1] [AC] [X,θ,T] [−] [X,θ,T] [SHIFT] [$x^2$] [F1] [2] [AC]

Because the Function Memory menu is already on the screen you do not have to enter $\boxed{\text{F}}$ MEM again, but can access the menu directly on the screen.

$$\boxed{\text{GRAPH}}\ \boxed{\text{F3}}\ \boxed{1}\ \boxed{\text{SHIFT}}\ \boxed{\leftarrow}\ \boxed{\text{GRAPH}}\ \boxed{\text{F3}}\ \boxed{2}\ \boxed{\text{EXE}}$$

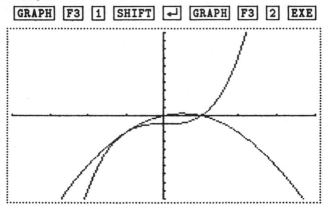

**Figure 6.39.** Stored functions: $y = x^3 - 1$ and $y = x - x^2$

Use some part of the zoom menu to find the point(s) of intersection.

**6.4.2  Piecewise Functions:** The Casio fx-7700G is able to graph piecewise defined functions. It is important to enter the domain (in interval notation) for each piece and enter each piece as part of a multistatement (see Section 6.3.2). The format is Graph y = function, [beginning of domain, ending of domain] for every piece you wish to graph. Set the range at the initial setting [-4.7, 4.7] by [-3.1, 3.1].

Now graph

$$f(x) = \begin{cases} \frac{1}{x} & \text{if } x < 1 \\ x - 1 & \text{if } 1 < x < 3 \\ \sin x & \text{if } x > 3 \end{cases}$$

Press: $\boxed{\text{GRAPH}}\ \boxed{1}\ \boxed{\div}\ \boxed{\text{X},\theta,\text{T}}\ \boxed{\text{SHIFT}}\ \boxed{,}\ \boxed{\text{ALPHA}}\ \boxed{\text{[}}\ \boxed{-5}\ \boxed{\text{SHIFT}}\ \boxed{,}\ \boxed{1}\ \boxed{\text{ALPHA}}\ \boxed{\text{]}}\ \boxed{\text{SHIFT}}$ $\boxed{\leftarrow}\ \boxed{\text{GRAPH}}\ \boxed{\text{X},\theta,\text{T}}\ \boxed{-}\ \boxed{1}\ \boxed{\text{SHIFT}}\ \boxed{,}\ \boxed{\text{ALPHA}}\ \boxed{\text{[}}\ \boxed{1}\ \boxed{\text{SHIFT}}\ \boxed{,}\ \boxed{3}\ \boxed{\text{ALPHA}}\ \boxed{\text{]}}$ $\boxed{\text{SHIFT}}\ \boxed{\leftarrow}\ \boxed{\text{GRAPH}}\ \boxed{\sin}\ \boxed{\text{X},\theta,\text{T}}\ \boxed{\text{SHIFT}}\ \boxed{,}\ \boxed{\text{ALPHA}}\ \boxed{\text{[}}\ \boxed{3}\ \boxed{\text{SHIFT}}\ \boxed{,}\ \boxed{5}\ \boxed{\text{ALPHA}}\ \boxed{\text{]}}$ $\boxed{\text{EXE}}$

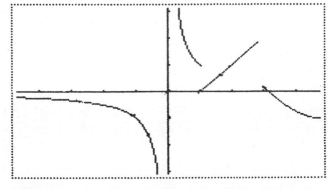

**Figure 6.40.** Piecewise Function (defined above)

Try using the trace function. Note that the left most value of the last function entered is where the pointer starts. The pointer can trace to each piece of the function. The left and right cursor keys move the pointer along one function. To change to another function, use the up and down cursor keys.

**6.4.3 Inequalities:** To graph inequalities, you must change the mode setting. In the text mode press [MODE] [SHIFT] to access the second mode window. Under graph type, press the [÷] key that corresponds to inequalities. Set the range at [-4.7, 4.7] scl 1 by [-5, 10] scl 2. When you press the [GRAPH] key the four inequality options appear at the bottom of the screen. Use the [F] keys to input the inequality you are graphing. Let's use this graphing capability to solve the system $y > x^2 - 4$ and $y < -x + 2$.

Press:     [GRAPH] [F1] [X,θ,T] [SHIFT] [$x^2$] [−] [4] [SHIFT] [EXE] [GRAPH] [F2] [−] [X,θ,T] [+]
           [2] [EXE]

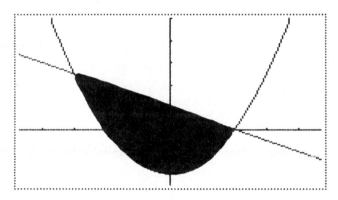

**Figure 6.41.** Inequalities: $y > x^2 - 4$ and $y < -x + 2$

The shaded area is the intersections of the two inequalities. This same graph can be used to solve $x^2 - 4 < -x + 2$. With each side graphed, simply identify the domain values of $x^2 - 4$ that have y-values below $-x + 2$. Using trace, we find this occurs when $-3.1 < x < 2$.

**6.4.4 Parametric Equations:** To graph a parametric equation, you must change to the parametric graphing mode by pressing [MODE] [SHIFT] [x]. Remember that you must define both the x and y components in a pair and that the independent variable in each must be T. Setting the range parameters requires both range screens (see description in 6.3.1). You must decide on using either radian or degree measure. Be sure that the mode setting agrees with the range parameter entries. For this example use radians—[SHIFT] [DRG] [F2] [EXE]. Graph $x = 5\sin T - 5\sin 2.5T$, $y = 5\cos T - 5\cos 2.5T$ with a range of [-15, 15] scl 5 by [-12, 12] scl 4; T, θ min 0; max $4\pi$; ptch $\pi/36$.

Enter:     [GRAPH] [5] [sin] [X,θ,T] [−] [5] [sin] [2] [.] [5] [X,θ,T] [SHIFT] [,] [5]
           [cos] [X,θ,T] [−] [5] [cos] [2] [.] [5] [X,θ,T] [EXE]

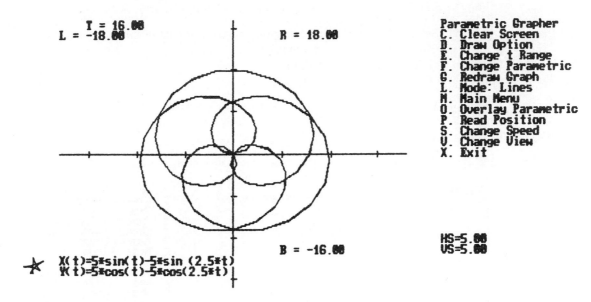

**Figure 6.42.** Parametric Equations: $x = 5 \sin T - 5 \sin 2.5I$ and $y = 5 \cos I - 5 \cos 2.5T$

As in function graphing, trace and zoom are available for exploring the graph. The TRACE feature lets you move the cursor along the equation one Tstep at a time. When you trace, the coordinate values of X, Y, and T are displayed at the bottom of the screen. As you trace along a parametric graph using $\boxed{\text{F1}}$, the values of X, Y and T are updated and displayed. The X and Y values are calculated from T. Scrolling is not possible on parametric curves. To see a section of the equations not displayed on the graph, you must change the RANGE variables. The ZOOM features work in parametric graphing as they do in function graphing.

## Application of Parametric Graphing

### An Example of Simulating Motion

*Problem*   Graph the position of a ball kicked from ground level at angle of 60° with an initial velocity of 40 ft/sec. (Ignore air resistance.) What is the maximum height, and when is it reached? How far away and when does the ball strike the ground?

*Solution*   If $v_o$ is the initial velocity and $\theta$ is the angle, then the horizontal component of the position of the ball as a function of time is described by

$$X(T) = Tv_o \cos \theta.$$

The vertical component of the position of the ball as a function of time is described by

$$Y(T) = -167T^2 + Tv_o \sin \theta.$$

In order to graph the equations,

(1)   Press:      **MODE** **SHIFT**. Select Parametric, Connected LIne, and Degree Mode.

(2)   Press:      **RANGE**. Set the RANGE variables appropriately for this problem.

Tmin = 0          Xmin = -5          Ymin= -5
Tmax = 2.5        Xmax = 50         Ymax = 20
Tstep = .02       Xscl = 5          Yscl = 5

(3)   Press:      **GRAPH**. Enter the expressions to define the parametric equation in terms of T.

$X_{1T} = 40T \cos 60$
$Y_{1T} = 40T \sin 60 - 16T^2$

(4)   Press:      **EXE** to graph the position of the ball as a function of time.

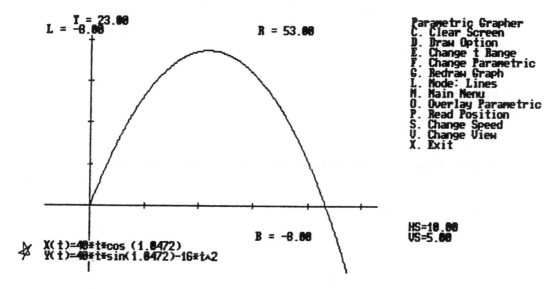

**Figure 6.43.** Parametric Applications: Position of Ball versus Time

(5)   Now press **TRACE** to explore the graph. When you press **TRACE**, the values for X, Y, and
      T are displayed at the bottom of the screen. These values change as you trace along the
      graph. Move the cursor along the path of the ball to investigate the maximum height, when
      it was reached, how far away and the time when the ball strikes the ground.

**6.4.5 Polar Equations:** To draw polar coordinate graphs you must change from the rectangular graph type to the polar graph type. This is done by pressing **MODE** **SHIFT** **−**. Also,
the angular measure must be in radians. To graph the built-in function $y = \sqrt[3]{\theta}$ press **GRAPH**
**SHIFT** **$\sqrt[3]{\phantom{x}}$** **EXE**.

*Note*   Built-in functions do not require the entry of $\theta$ or the setting of range parameters—they
have preselected ranges.

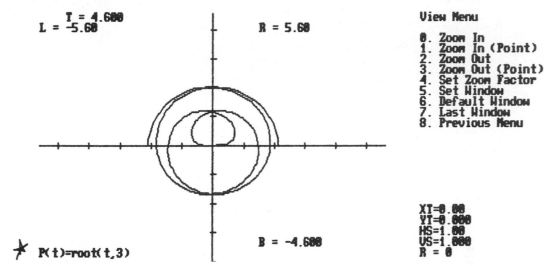

**Figure 6.44.** Polar Equation: $r = \sqrt[3]{\theta}$

To graph user-generated graphs the range parameters must be set. Both range screens must be used (see description in Section 6.3.1).

To graph $r = 3\sin 2\theta$, we first set the ranges at [-3.6, 3.6] scl 1 by [-2.6, 2.6] scl 1; T, $\theta$ min $-\pi$; max $\pi$; ptch $\pi/36$ (note that the range values accept multiples of $\pi$).

Now enter the function.

Press:  GRPH  3  sin  2  X,$\theta$,T  EXE

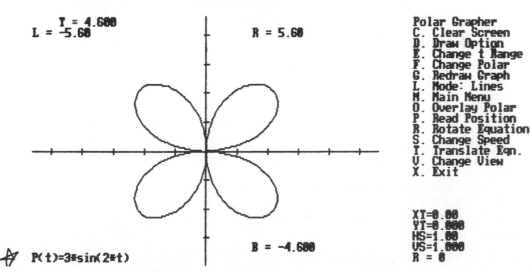

**Figure 6.45.** Polar Equation: $r = 3\sin 2\theta$

As with rectangular and parametric graphing, the trace and zoom functions can be used with polar graphs in the same way as with parametric graphing. The scrolling feature can not be used.

# Chapter 7

## Calculating and Graphing with the Sharp EL-5200

The Sharp *EL-5200* graphics calculator is a versatile, easy-to-use tool that can be useful in many of your mathematics courses. In addition to the usual abilities of a programmable scientific calculator the *EL-5200* graphs functions, computes one- and two-variable statistics, and does matrix operations. Like any good tool, you will need to care for and protect the *EL-5200*. If you heed the cautions in Section 7.1, your *EL-5200* should continue to be a mathematical assistant to you for a long time.

### 7.1  How Best to Protect Your Sharp EL-5200

(1) *Place the calculator on a flat surface when using it.* **Do not** fold over the cover containing the right-hand keyboard. If you fold over the cover it is very likely that you will damage the cables between the right and left sides of the calculator.

(2) *Press the keys on the right side of the calculator with your finger or soft objects (for example, a pencil eraser).* **Do not** press the keys on the right side with the sharpened end of a pencil or with a ball point pen.

(3) *Like all calculators your Sharp will not continue to operate correctly if it is bent, bumped, or dropped.* For example, **do not** carry the *El-5200* in your pants or slacks pocket. Put it in a protected place if you carry it in your backpack.

(4) *Keep your calculator cool and protect it from dust.* If it needs cleaning, use a soft, **dry** cloth.

(5) *All of the keys may become inoperative if the calculator is subjected to a loud noise or severe shock while in use*—at least that is what is stated in the *Owner's Manual*. If this should happen consult the page whose heading is **Operational Notes** in the *Owner's Manual*.

### 7.2  Some Basics

This primer is intended to help you learn to draw graphs of functions using the *El-5200*. If you become interested in the many other things the *EL-5200* can do, read the *Owner's Manual*. The last part of Chapter 2 in that *Manual* tells you how to operate on matrices. Chapter 3 discusses statistics, and Chapter 6 shows how to program the *EL-5200*.

The *EL-5200* has four operating modes: [AER-I], [AER-II], [COMP], and [STAT]. The first two modes are the ones you would use if you wanted to program your calculator. The [COMP] mode is the numeric computation and graphing mode and is the one that will be used throughout this chapter. The [STAT] mode is the one used for statistics. Since the [COMP] mode will be used throughout this discussion, put your *EL-5200* in that mode now. On the far left side of the calculator you will see a sliding switch and a diagram that indicates the [COMP] mode is the third position from the top; put the sliding switch in that position.

The *EL-5200* has two keyboards. The left-hand keyboard includes the keys typically found on a scientific calculator; this keyboard is used for most numerical computations. For this reason the left-hand keyboard is called the *numeric keyboard*. The right-hand keyboard has on it the keys that control the calculator's graphics; it is called the *graphics keyboard*.

The top row of the numeric (left-hand) keyboard consists of the [ON], [OFF], and cursor keys ([▽], [△], [◁], [▷]). The last four keys on this row also have other roles when they are used with the left-most and most important key on the second row of this keyboard, the [2ndF] key. The abbreviation [2ndF] stands for "second function." Notice that the [2ndF] key is gold in color and that scattered over the numeric and graphics keyboards are other keys with gold symbols on some part of them. For example, on the top row of the numeric keyboard the [▽] key has the gold symbol [$T \triangleright G \triangleright D$] above it. When the two keys [2ndF] and [▽] are pressed in this order, the screen "toggles" through the **T**ext, **G**raphics, and **D**ata displays. There is also a [2ndF] key on the upper right-hand corner of the graphics keyboard; pressing it or the one on the numeric keyboard has the same effect. Other important keys on the second row of the numeric keyboard are the [FSE], [PB], [DEL], and [CL] keys; these keys will be discussed a little later.

The remaining rows of the numeric keyboard are familiar if you have used a scientific calculator before. If you have not used a scientific calculator previously, the functions of many of these keys will be introduced to you as you study from the accompanying textbook; these keys will not be discussed extensively here.

## 7.3 Getting Started

Put your calculator in the [COMP] mode; then turn it on by pressing the [ON] key at the top left corner of the numeric keyboard. When you do, you should be able to distinguish the "active" area of the screen; if not adjust the screen contrast. To lower the contrast alternately press the [SHIFT] key and the [∨] key; to raise the contrast alternately press the [SHIFT] key and the [∧] key.

Now that your calculator is on, you will see two rows of small symbols also displayed under the active portion of the screen. In the top row of symbols only one symbol is of consequence here; it is the [♩] symbol. If this symbol is lighted then whenever you touch a key on either keyboard you will hear a beeping sound. To turn the beeper—or click—on and off press [2nd] [♩]; the [♩] is the second function of the [0] key on the graphics keyboard.

If the **2ndF** symbol is lighted, then the [2ndF] key has been depressed, and if the **SHIFT** symbol is lighted then the [SHIFT] key has been depressed. A lighted **HYP** indicates a hyperbolic function has been chosen. The next symbol on the second row of symbols is important; the part of it that is lighted should be **RAD**. If **RAD** is not lighted, then press [2ndF] [DRG] (atop the [FSE] key on the second row of the numeric keyboard) until it is. None of the symbols **FIX**, **SCI**, or **ENG** should presently be lighted. If any of these three symbols is lighted press [FSE] on the second row of the numeric keyboard until none of them are lighted. When lighted the four symbols ←, ↑, ↓, and → indicate that you can use the cursor keys on the top row of the numeric keyboard to move in that direction. The only symbol that must be lighted at present is **RAD**; having this symbol lighted indicates that radians are being used for measuring angles; this ensures, for example, appropriate calculation of $\sin(2\pi/3)$. The other units of angle measure that you can use are degrees (**DEG**) and grads (**GRAD**); you may want to experiment with these but always remember to return you calculator to radian measure.

Anytime that you are through using your *EL-5200*, press the [OFF] key. If you forget to do this, it will turn itself off automatically when it has not been used for about 10 minutes. If the calculator happens to turn itself off, simply press the [ON] key to reactivate it.

## 7.4  Error Messages

Occasionally everyone makes an error when using a calculator. The six error messages that the *EL-5200* uses are written on the back of the numeric keyboard; some causes of each of these errors are described in Appendix D of the *Owner's Manual*. The error message that you are most likely to see is "**ERROR 1**;" this is a syntax error. Figure 7.1 shows some keying sequences that will result in this error message.

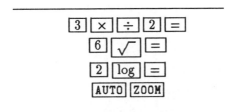

**Figure 7.1.** Examples of ERROR 1 keying sequence mistakes.

If you should make an error, you should try to correct it *instead of pressing* $\boxed{\text{CL}}$ *and starting over again*. To do this press the playback ($\boxed{\text{PB}}$) key on the top row of the numeric keyboard. When you do this, the cursor will begin to flash at the place in the last line executed where the calculator found the incorrect instruction. For example if you should mistakenly enter $\boxed{7}\boxed{(\,-\,)}\boxed{3.4786}\boxed{=}$ you will receive an **ERROR 1** message since you pressed the $\boxed{(\,-\,)}$ key (the negative key) instead of the subtraction key, $\boxed{-}$. When you press $\boxed{\text{PB}}$, the cursor will begin to flash on the $\boxed{(\,-\,)}$ instruction. To correct this mistake simply press the $\boxed{-}$ key; then press the $\boxed{=}$ key again, the correct result (3.5214) will appear on the next line. Sometimes you may end up almost completely reentering the instructions on a line (for example, when you accidentally enter $\boxed{6}\boxed{\sqrt{\ }}\boxed{=}$), but you will know where you made your initial mistake if you press the $\boxed{\text{PB}}$ key and can often determine what needs to be corrected.

## 7.5  Using the EL-5200 Editor

A table will sometimes help you better understand a function, discover a pattern, or validate your predictions. Making a table of the values of a function is really very easy with the *EL-5200*. For example, suppose that you wanted to make a table of values for the function $f(x) = \ln(x)$. First determine the values of $x$ at which you want to compute $f$; let's take some easy ones: $1, 2, 3, 4, 5, \cdots$. Enter the function using for $x$ the first value you want to compute; in this case $1$. The screen on your *EL-5200* should look like Figure 7.2 if you clear the screen before beginning your table.

$$\emptyset .$$

$$\text{LN } 1 =$$

$$\emptyset .$$

**Figure 7.2.** Computation of $\ln(1)$.

Enter the value of $f(1)$ into your table before continuing. Now you are ready to compute $f(2)$. Press the PB key on your calculator. Using the ◁ move the cursor until it is over the "1" in the expression on the screen; when it is press the 2 key and then the = key. The value of $f(2)$ will appear on the screen on the next line; write down that value. Continue in this way until you have computed all of the values of $f(x) = \ln(x)$ that you need.

If you want to change the function whose values you are computing, that is easy too. Once again press the PB key. When the cursor is flashing at the end of the line, simply press the left-hand cursor key until the cursor is over **ln**. Then press the key corresponding to the new function whose values you want to compute. You can then proceed to make a table of values for the new function in the same way that you did for $f(x) = \ln(x)$.

When editing a line of instructions requiring more space in some place, first position the cursor at the place where you need that space and press 2ndF INS; this will give you a place for an additional keystroke. If you need to delete a digit or an instruction, position the cursor at the instruction or keystroke that you want to delete and press DEL.

## 7.6  EL-5200 Graphics

Drawing graphs of functions using your *EL-5200* is very easy. The *EL-5200* has three display windows: the **T**ext, **G**raphics, and **D**ata displays. You "toggle" through these displays by pressing the 2ndF $T \triangleright G \triangleright D$ keys; you will quickly learn to recognize which display you are viewing. If you are looking at the text display, pressing 2ndF $T \triangleright G \triangleright D$ once will toggle you to the graphics display. The next time you press 2ndF $T \triangleright G \triangleright D$, you will be viewing the data display. A final use of 2ndF $T \triangleright G \triangleright D$ will bring you back around to the text display. There is another way to go directly back and forth between the text and graphics displays that will be discussed when techniques for displaying two functions simultaneously on the graphics display are discussed.

**The Viewing Rectangle.** The *EL-5200* has a *default viewing rectangle* whose $x$-range is $[-4.7, 4.8]$ and whose $y$-range is $[-1.5, 1.6]$. To obtain this default viewing rectangle press the RANGE key on the top row of the *graphics keyboard*. Then press 2ndF CA (the CA key is on the second row of the *numeric keyboard*). When you do this, the calculator screen should look like Figure 7.3.

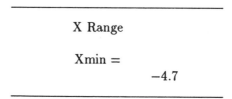

**Figure 7.3.** Display of the Range using the EL-5200 default viewing rectangle.

You can scroll through the range values by using the ▽ and ∧ cursor keys. While the Range values are displayed, you can change them. Simply move the cursor to the value that you want to change, type in the new value, and press the = key. A good beginning point for most of the functions graphed in the accompanying textbook is the standard viewing rectangle of $[-10, 10]$ by $[-10, 10]$. This translates to the Sharp Range of

$$\text{Xmin} = -10 \quad \text{Xmax} = 10 \quad \text{Xscale} = 1 \quad \text{Ymin} = -10 \quad \text{Ymax} = 10 \quad \text{Yscale} = 1.$$

The values of Xdot and Ydot will be set automatically when you set the six values listed above (see *Owner's Manual*, pp. 90–91). When you have set the Range values to the ones that you want to use, press the $\boxed{\text{RANGE}}$ key; either your calculator will return to the text display or the cursor will jump down to the line **Ymin =** . If the cursor goes to the latter place, press the $\boxed{\text{RANGE}}$ key again, and the text display will appear.

**Drawing a Graph.** To graph a function you will use familiar keys on the *numeric keyboard* and three keys on the *graphics keyboard:* $\boxed{\text{GRAPH}}$, $\boxed{\text{DRAW}}$, and $\boxed{\text{X}}$. You will need to use $X$ as the name of the variable in any function that you draw because it is the only variable name that the *EL-5200* "understands." Suppose you want to graph $f(x) = x^2$. This keystroke sequence will graph $f$:

$$\boxed{\text{GRAPH}}\;\boxed{\text{X}}\;\boxed{x^2}\;\boxed{\text{DRAW}}\;.$$

When you press $\boxed{\text{DRAW}}$, the screen will display COMPUTING and the $\leftarrow$ and $\rightarrow$ symbols will flash on the bottom line of the display. After a few seconds, the graph will appear; the amount of time required to graph a function varies and depends upon the complexity of the function being graphed. Figure 7.4 shows the graph of $f(x) = x^2$ as it will appear on your screen if you use the standard viewing rectangle of $[-10, 10]$ by $[-10, 10]$.

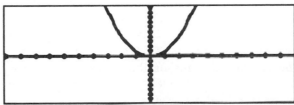

**Figure 7.4.** Sharp *EL-5200* graph of $f(x) = x^2$ in the viewing rectangle $[-10, 10]$ by $[-10, 10]$.

**Editing the Function Graphed.** To change the function $f$ to $f(x) = 3x^2 + x - 2$, press the $\boxed{\text{PB}}$ key and then enter these keying sequences:

(1) Using the $\boxed{\triangleleft}$ key move the cursor so that it is on $X$. Press $\boxed{\text{2ndF}}\;\boxed{\text{INS}}$ and then $\boxed{3}$. The function has now been changed to $f(x) = 3x^2$. On the *EL-5200* you need to key only $\boxed{3}\boxed{\text{X}}$ to get the $3x$ instead of keying $\boxed{3}\boxed{\times}\boxed{\text{X}}$.

(2) Using the $\boxed{\triangleright}$ move the cursor so that it is on the "D" in DRAW and press the keys $\boxed{+}\boxed{\text{X}}\boxed{-}\boxed{2}$. Now the function is $f(x) = 3x^2 + x - 2$. Press $\boxed{\text{DRAW}}$, and the graph of the new function will be computed and then shown on the screen.

If the new function to be graphed is very different from the one previously graphed, it is probably easier simply not to edit the existing function but to enter in the new function "from scratch." To do this press $\boxed{\text{PB}}$, and then press the $\boxed{\triangledown}$ key until you have moved down below the last line of instructions currently in the text window; at that point the cursor will cease being a flashing block (■) and will become the underline ( _ ) character. Now simply press $\boxed{\text{GRAPH}}$, enter the function, and then press $\boxed{\text{DRAW}}$.

**Erasing the Graphics Display.** Because the graphics display was not erased (i.e., cleared) after the function $f(x) = x^2$ was graphed, two functions are now showing in the graphics window. To see only the graph of $f(x) = 3x^2 + x - 2$, first press $\boxed{\text{2ndF}}\;\boxed{\text{G.CL}}$; the $\boxed{\text{G.CL}}$ key is on the top row of the numeric keyboard. When you press these two keys, the graphics screen will be cleared and only the Cartesian axes will be shown. Now, press $\boxed{\text{PB}}$; the cursor will be flashing over the "G" in G.CL. Use the $\boxed{\wedge}$ key to move the cursor to the end of this line:

$$\text{GRAPH } 3X^2 + X - 2 \,\text{DRAW}$$

When you have done this press the $\boxed{\text{DRAW}}$ key, and the function $f(x) = 3x^2 + x - 2$ will be redrawn in the graphics window.

**Zoom-In and Zoom-Out.** To see more of the graph of a function, it is necessary to *zoom-out*; to see the local behavior more clearly, it is necessary to *zoom-in*. Both of these features are best accomplished using the $\boxed{\text{AUTO}}$ and $\boxed{\text{ZOOM}}$ keys; the $\boxed{\text{ZOOM}}$ key is on the top row of the *graphics keyboard* while the $\boxed{\text{AUTO}}$ key is on the second row of that keyboard. To see how to zoom-in and to zoom-out, begin with the graph of a function on the graphics screen. For example, use the graph of the function $f(x) = 3x^2 + x - 2$ in the viewing rectangle $[-10, 10]$ by $[-10, 10]$ that was just discussed. To see this function in the viewing rectangle $[-20, 20]$ by $[-20, 20]$, press $\boxed{\text{AUTO}}\boxed{.}\boxed{5}\boxed{\text{ZOOM}}$. In a short while the graph of the function $f$ will reappear except that this time it will seem "skinnier" than it was before. To check that the range on both the $x$- and $y$-axes is now from $-20$ to $20$, press the $\boxed{\text{RANGE}}$ key and use the $\boxed{\triangledown}$ and $\boxed{\wedge}$ keys to check the range values. However, **do not** change any of the range values; **if you do then the text display will be completed cleared and you will have to begin all over by first entering a function to be graphed.**

When you are satisfied that the zoom-out feature works as described, before proceeding notice the value used between the instructions **AUTO** and **ZOOM**—it was $0.5$. This value doubled both the $x$- and $y$-ranges. The way to determine the value you use is this: take the reciprocal of the effect that you want to achieve. For example, if you want to double the ranges (a factor of $2$—a zoom-out), then enter in the value $1/2$ (the reciprocal of $2$); if you want to cut the ranges by $1/3$ (i.e., zoom-in) enter in $3$ (the reciprocal of $1/3$). Values less than $1$ will produce a **zoom-out**; values greater than $1$ will produce a **zoom-in**. Thus, since in the example, a zoom-out that doubled the ranges was used, to return to the original graph in the $[-10, 10]$ by $[-10, 10]$ viewing rectangle you want to cut the size of the present ranges by half: enter $\boxed{\text{AUTO}}\boxed{2}\boxed{\text{ZOOM}}$. The original graph will appear; check the range values to see that this is true. **Remember** not to change any of the range values, or you will need to start all over.

The amount that the $x$- and $y$-ranges are altered using the $\boxed{\text{AUTO}}$ and $\boxed{\text{ZOOM}}$ keys can be controlled individually; that is, for example, you can zoom-in on the $x$-axis while simultaneously zooming-out on the $y$-axis. In the present example, this very action would probably improve the graph of $f$. To make the viewing rectangle $[-2.5, 2.5]$ by $[-20, 20]$ from $[-10, 10]$ by $[-10, 10]$ it is necessary to make the $x$-range one-fourth its present size (i.e., to use a value of $4$) and to double the $y$-range (i.e., to use a value of $0.5$). To do this key in $\boxed{\text{AUTO}}\boxed{4}\boxed{,}\boxed{.}\boxed{5}\boxed{\text{ZOOM}}$; look at the graph and then check the range.

**Two Functions and Zooming.** There are times when you will want to work simultaneously with the graphs of two functions and to zoom-in or zoom-out on both of these functions. To illustrate how to do this, set the viewing rectangle to $[-10, 10]$ by $[-10, 10]$; let the first function be $f(x) = 3x^2 + x - 2$; and let the second function be:

$$g(x) = x^3.$$

(To graph $g$ use the keying sequence: $\boxed{\text{GRAPH}}\boxed{\text{X}}\boxed{\text{Y}^X}\boxed{3}\boxed{\text{DRAW}}$.) To zoom-in or zoom-out efficiently on both graphs, it is convenient to have them in two adjacent lines of the text display. If this is not the case in your calculator, return to the text display, and once there, press $\boxed{\text{2ndF}}$ $\boxed{\text{CA}}$; this will clear the text workspace. Now, you will have to reenter both the function $f$ and the function $g$ by entering *two consecutive* GRAPH/DRAW statements. When both functions have been graphed your graphics display should look like that in Figure 7.5.

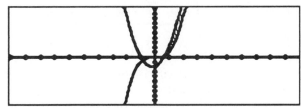

**Figure 7.5.** Graph of $f(x) = 3x^2 + x - 2$ and $g(x) = x^3$ in the viewing rectangle $[-10, 10]$ by $[-10, 10]$.

A better viewing rectangle for both of these functions is $[-4, 4]$ by $[-4, 4]$. To change to this viewing rectangle a value of 2.5 is needed for both the $x$-axis and the $y$-axis. To enter this press $\boxed{\text{AUTO}}$ $\boxed{\text{2.5}}$ $\boxed{\text{ZOOM}}$. When the graphics display reappears, only the graph of $g(x) = x^3$ will be on the screen; the reason for this is that when the $\boxed{\text{ZOOM}}$ key is pressed *only* the last graph drawn is redrawn. To return the graph of $f(x) = 3x^2 + x - 2$ to the screen, do this: (a) press $\boxed{\text{PB}}$, (b) move the cursor up using the $\boxed{\wedge}$ key until the cursor is flashing over the beginning of **DRAW** in the line

$$\text{GRAPH } 3X^2 + X - 2 \text{ DRAW},$$

and (c) press the $\boxed{\text{DRAW}}$ key. You can continue zooming in and out by pressing the $\boxed{\text{PB}}$ key and editing the line **AUTO 2.5 ZOOM**. The next time only $f$ will be redrawn when you press $\boxed{\text{ZOOM}}$ since it was the last graph drawn; so you will need to redraw $g$ using the instructions just given.

**Finding the Coordinates of a Point.** In mathematics we often want to know the $x$- and $y$-coordinates of a point on the graph of a function. For example, we want to know where the graphs of two functions intersect each other, where a function has a zero (i.e., crosses the $x$-axis), and where a function achieves its local minimum and maximum values. Until recently most of the values were found by using algebraic and analytic techniques. However, using the *EL-5200* and its zoom-in feature it is possible to approximate either or both of the $x$- and $y$-coordinates of a point of interest.

To show how to do this we will continue using the functions $f$ and $g$ as defined earlier and will assume that the present viewing rectangle is $[-4, 4]$ by $[-4, 4]$. In this viewing rectangle it can be seen that the two functions intersect each other twice; they intersect one time with an $x$-coordinate that is somewhere between $-1$ and $0$ and another time with an $x$-coordinate that is between $0$ and $1$. The rest of this example shows how to approximate the $x$- and $y$-coordinates (accurate to $0.01$) of the first point of intersection.

With the graphs of the two functions displayed press the $\boxed{\triangleright}$ cursor key. When you press this key the screen will appear as in Figure 7.6; if you will look carefully you will see that a small dot is flashing at the left end of the last graph that was drawn. The $x$-coordinate value displayed on the bottom of the screen is that of the point that is flashing; the flashing indicates that the tracing cursor has been activated. In this case both graphs have the same first *on-screen* $x$ value. This is coincidental.

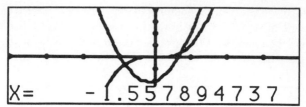

**Figure 7.6.** Graph of $f$ and $g$ in the viewing rectangle $[-4, 4]$ by $[-4, 4]$ with the tracing cursor activated.

You can move the cursor along the curve from *left to right* by continuing to press the $\boxed{\triangleright}$ key; you can move the cursor from *right to left* by pressing the $\boxed{\triangleleft}$ key. As long as you remain within the present viewing rectangle both graphs will continue to remain on the display screen. However, if by pressing either of these keys you cause the curve you are tracing to move, then the other graph will *disappear from the screen* because you have changed the viewing rectangle slightly! You can recover the missing graph by using the technique outlined in the preceding part of this section.

Before finding the point of intersection of $f$ and $g$ that has $x$-coordinate between $-1$ and $0$, press $\boxed{\text{2ndF}}$ $\boxed{X \leftrightarrow Y}$. When you do the $y$-coordinate of the point that is presently flashing will be displayed on the screen. Pressing this keying sequence again will cause the value of the $y$-coordinate to disappear though the point will continue to flash. Pressing this keying sequence a third time will return the $x$-coordinate to the bottom of the screen. To prepare for a zoom-in about the third quadrant point of intersection of $f$ and $g$, press $\boxed{\text{PB}}$ $\boxed{\triangledown}$; this should take you just below the two lines of instructions that graphed $f$ and $g$. Now press $\boxed{\text{AUTO}}$ $\boxed{4}$, and then return to the graphics window by pressing $\boxed{\text{2nd}}$ $\boxed{T \triangleright G \triangleright D}$. Next press $\boxed{\triangleright}$ repeatedly until the flashing point is approximately at the third-quadrant point of intersection. When it is, press $\boxed{\text{ZOOM}}$. When the zoom-in is completed, recover the other graph. To zoom-in again, use $\boxed{\text{PB}}$ followed by repeated use of $\boxed{\triangleleft}$ to place the flashing cursor on the $z$ in zoom. Then press $\boxed{\text{2nd}}$ $\boxed{T \triangleright G \triangleright D}$ followed by repeated use of $\boxed{\triangleright}$ to trace to the point of intersection. Once there, press $\boxed{\text{ZOOM}}$ with the graphics window still showing. As usual, you will have to recover the second graph. Continue to do this until when you move the cursor to the point of intersection the values of the $x$- and $y$-coordinates of the point of intersection differ from those next to it by less than $0.005$. When this occurs, the approximation of the coordinates of the point of intersection are accurate to $0.01$. In this case, you should obtain the ordered pair $(-0.86, -0.64)$.

The techniques for finding the $x$- and $y$-coordinates of a point have been illustrated by finding a point of intersection of two graphs. The same techniques can be used whenever you need to approximate graphically the $x$- or $y$-coordinates of any point.

**7.6.1  Polar Graphs:** To graph polar functions, i.e. functions of the form $r = f(\theta)$, you will need to enter a simple program into your *EL-5200* in the AER-II programming space. First, place the slide switch on the left side of the *EL-5200* beside the second mark from the top. Next press the $\boxed{\text{PRO}}$ key and enter in the title "polar." Then press $\boxed{\text{ENT}}$; the screen will clear except that in the upper left-hand corner will be "M:". Following this you will enter the program.

Suppose that you wanted to graph $r = 1 + 2\cos(t)$ in the range $0 \leq t \leq 360$. Then the program you would enter into "polar" would be this one:

M: $\boxed{\text{RANGE}}$-1,4,1,-2,2,1 ⊔$t = 0$⊔ $\boxed{\hookleftarrow}$ $r = 1 + 2\cos(t)$⊔$x = r\cos(t)$⊔$y = r\sin(t)$⊔ $\boxed{\text{PLOT}}$x,y $\boxed{\text{DRAW}}$ ⊔$t >=$ 360 $\boxed{\blacksquare Y \blacksquare}$ [ $t = t$ $\boxed{\blacksquare}$ $\boxed{]}$ $\boxed{\blacksquare N \blacksquare}$ [ $t = t + 5$ $\boxed{\hookleftarrow}$ $\boxed{]}$

Once you have entered this program press $\boxed{\text{ENT}}$. To run "polar" return your calculator to the COMP mode. Press $\boxed{\text{PRO}}$ and then $\boxed{\text{COMP}}$. The graph of the polar function will appear on the screen. To change the function simply change $r = 1 + 2\cos(t)$ to whatever function you want to graph next. You may also need to change the range values (including those for $t$) if you do not get a complete graph using the range in the program given above.

**7.6.2 Parametric Graphs:** A program similar to the one given above will graph parametric equations. For example here is the program you would enter in in the AER-II mode for $x = t + 3$ and $y = t^2 - 2$ in the viewing rectangle $[-3, 9]$ by $[-3, 8]$.

M: $\boxed{\text{RANGE}}$-3,9,1,-3,8,1 $\sqcup t = -3$ $\boxed{\hookleftarrow}$ $x = t + 3\sqcup y = t^2 - 2\sqcup$ $\boxed{\text{PLOT}}$x,y$\boxed{\text{DRAW}}$ $\sqcup t >= 3$ $\boxed{-Y \rightarrow}$ $\boxed{[}$ $t = t$ $\boxed{\blacktriangle}$ $\boxed{]}$ $\boxed{-N \rightarrow}$ $\boxed{[}$ $t = t + 0.1$ $\boxed{\hookleftarrow}$ $\boxed{]}$

Returning to the COMP mode and proceeding as for polar graphs will give you the graph of the parabola having these parametric equations.

## 7.7 Conclusion

The Sharp *EL-5200* is a versatile graphics calculator that can do much more than has been discussed here. However, this chapter should begin to make you comfortable while using the *EL-5200*, particularly while solving the exercises and problems posed in the accompanying textbook. If you need or want additional information about the capabilities of the *EL-5200*, we suggest that you read the *Owner's Manual*. This manual is sometimes not very easy to understand, but if you will read only those parts currently of interest to you, its meanings should become clear.

# Chapter 8

## Computing and Graphing with the HP-28S

This is an introduction to the use of the HP-28S as a graphing calculator. No previous knowledge of the HP-28S is assumed, and much of what is discussed can be found in the *Owner's Manual*. Section 8.1 is a quick lesson on the HP-28S as a scientific calculator. It includes an introduction to arithmetic and algebraic expressions as well as a discussion of automatic root-finding. Section 8.2 covers the built-in graphing commands. Section 8.3 gives enough of an introduction to programming and directory management to create a customized graphing directory.

### 8.1 The HP-28S as a Scientific Calculator

Every key, except the top six on the right keyboard, has two functions—a primary function and an alternate function, which appears written in red immediately above the key and is activated by the red shift key (there is no red shift lock). To turn the calculator on, press $\boxed{\text{ON}}$ on the right keyboard. To turn it off, use the alternate function of ON. That is, press the red key followed by the ON key, which we will write as $\blacksquare\,\boxed{\text{OFF}}$. The ON key also has written below it in white ATTN, which means ATTENTION (!). This is because the ON key also behaves something like CONTROL C on a computer. It is the key to press when in trouble or doubt. You use it to

(1) stop execution of any program,

(2) clear error messages from screen,

(3) clear the "command line" (more on that later),

(4) clear the screen of a graphic display and return to normal text display.

Now that you have turned the machine on, let's add 5 and 7. You'll probably get a beep and the error message "+ Error: Too Few Arguments." That's because you may have pressed 5 followed by +, after which you hoped to press 7 and then search for =. We need to try a different approach. The HP-28S has two modes of performing calculations. The first uses the "stack," and the second is a direct entry mode.

**8.1.1 Stack Mode Using Reverse Polish Notation:** The HP uses postfix order of operations, taking the most recent entries and operating on them. The HP expects arguments to be entered before operators. This approach is called *reverse Polish notation*, or RPN.

```
4:                         4:
3:                         3:
2:                         2:
1:                         1:                    5
```

Figure 8.1                              Figure 8.2

To compute $5 + 7$, first clear the stack by pressing ■CLEAR (the red key followed by 0). Then, if there is a menu across the bottom of your screen, press ◀↕▶. The screen of your HP should now look like Figure 8.1. Now press 5 followed by ENTER (see Figure 8.2), then 7 again followed by ENTER (see Figure 8.3), then + (see Figure 8.4). In this process you will see first the number 5 appear at the far left of the bottom line of the screen. When you press ENTER the number 5 is placed on the *stack*, which is the general work area of the calculator. The levels of the stack are numbered 1:, 2:, 3: etc. For all intents, the stack is infinitely deep. When you first pressed 5, the number 5 did not go on the stack; instead it appeared on the *command line* below the stack on the bottom of the screen. ENTER empties the command line and enters it on the stack. When you press 7 ENTER, the number 7 goes to level 1 of the stack pushing 5 to level 2. The key + then adds what is on level 1 to what is on level 2 and places it at level 1. To drop what is on level 1 of the stack press DROP; to clear the entire stack press ■CLEAR.

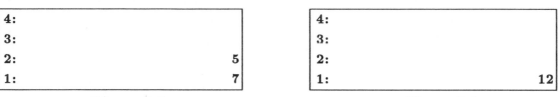

```
4:                         4:
3:                         3:
2:                    5     2:
1:                    7     1:                   12
```

Figure 8.3                              Figure 8.4

All other standard arithmetic operations work in the same way. To compute 2/3, press 2 ENTER 3 ENTER ÷ or 2 ENTER 3 ÷. To compute $2^3$, press 2 ENTER 3 ■ ∧ (red function of ×).

Now suppose we want to take the sine or natural log of a number. You'll find no SIN key. That is because the HP-28S is menu driven. Press TRIG (second row, right side). You will see six black boxes across the bottom of the screen: SIN, ASIN, etc. This is a menu of commands for the top six keys of the right keyboard. To compute the sine of 58, press 58 ENTER, and then press the key immediately below SIN on the TRIG menu. The TRIG menu actually has more commands. To see the next six menu items press NEXT (second row, right side). Press NEXT again to see the next six. Press NEXT again and you have gone all the way around the TRIG menu back to the first six items. The exponential and natural log functions are on the LOGS menu (red function of TRIG), along with log base 10, LOG, and its inverse, ALOG.

When you computed the sine of 58, you might well ask whether it was the sine of 58 degrees or of 58 radians. This depends on what "mode" the calculator was in. Press ■MODE (second row, right side). Then press the key under DEG on the MODE menu if you want degrees mode or RAD if you want radians (a small ($2\pi$) appears on the top of screen if you are in radian mode). As long as we're on the MODE menu, let's look at the possible ways numbers are displayed. If you press the key under STD (standard), numbers will be displayed to 12 digits, rounded, with terminating zeroes deleted. In particular, integers are displayed without decimal points. If you press 3 followed by the key under FIX, numbers will be displayed with 3 digits to the right of the decimal point (even if these digits are zeroes; so the integer 4 is displayed as 4.000). If the number is too small or too big, the display reverts to standard scientific notation, with 3 digits to the right of the decimal point. If you press $\boxed{7}$ followed by the key under SCI, all numbers will be displayed in scientific notation with 7 digits to the right of the decimal point.

One last thing about modes: Suppose you want to find $\sin(\pi/6)$. First make sure you are in radian mode (see preceding paragraph). Then press $\blacksquare\,\boxed{\pi}$ (red function of $\boxed{\cdot}$ in last row, right side) followed by $\boxed{\text{ENTER}}$. You will see the symbol $'\pi'$ on the stack. If you press $\boxed{6}\ \boxed{\div}$ you'll get $'\pi/6'$. If you press $\boxed{\text{SIN}}$ on the TRIG menu you get $'\text{SIN}(\pi/6)'$. In general, the HP-28S treats $\pi$ as a symbol for the actual (not approximate) value of $\pi$. To get a numerical value for $'\pi'$ or $'\text{SIN}(\pi/6)'$ press $\blacksquare\,\boxed{\to\ \text{NUM}}$ (red function of $\boxed{\text{EVAL}}$). If you want $\pi$ treated always as the number 3.14159265359, then type:

$$\boxed{35}\ \boxed{\text{ENTER}}\ \boxed{\text{C}}\ \boxed{\text{F}}\ \boxed{\text{ENTER}}.$$

You will have to type CF using the alphabetic keyboard on the left. This clears "flag" 35. Try it, and then press $\blacksquare\,\boxed{\pi}\ \boxed{\text{ENTER}}$ to see what happens. You'll probably want to leave your calculator this way, but if you want $\pi$ treated as a symbol again, key in $\boxed{35}\ \boxed{\text{ENTER}}\ \boxed{\text{S}}\ \boxed{\text{F}}\ \boxed{\text{ENTER}}$.

At this point, you should be able to use the HP-28S as a standard scientific calculator, except we haven't shown you how to store constants. Numbers are kept on the stack unless you clear them, and there are some stack commands to access items on the stack. The most useful are SWAP (red function, third row, right side) which swaps level 1 and 2, and ROLL (red function of DROP) which requires some explanation.

First, clear the stack by pressing $\blacksquare\,\boxed{\text{CLEAR}}$. Then place the first four multiples of 10 on the stack by pressing

$$\boxed{40}\ \boxed{\text{ENTER}}\ \boxed{30}\ \boxed{\text{ENTER}}\ \boxed{20}\ \boxed{\text{ENTER}}\ \boxed{10}\ \boxed{\text{ENTER}}.$$

Your screen should look like Figure 8.5. Now press $\boxed{4}\ \blacksquare\,\boxed{\text{ROLL}}$. Your screen should now look like Figure 8.6.

| 4: | 40 |
|----|----|
| 3: | 30 |
| 2: | 20 |
| 1: | 10 |

**Figure 8.5**

| 4: | 30 |
|----|----|
| 3: | 20 |
| 2: | 10 |
| 1: | 40 |

**Figure 8.6**

Notice that the entry, 40, in the level selected for the ROLL command, level 4, has been dropped to level 1 and that the other entries have been "rolled" back up the stack. You may wish to experiment with the ROLL command to gain control of this useful feature.

The USER memory allows you to store values under variable names that then appear on the user menu. We'll discuss this storage method later.

In the meantime, there are a few other things you need to know. If you make a mistake while typing something on the command line, you can always press $\boxed{\text{ON}}$ and start over. You can also press the backspace key $\boxed{\Leftarrow}$ (third row, right side) to back over what you've written and erase it in the process. Finally you can actually edit the command line, moving the cursor over characters and overwriting or inserting or deleting. To do this press $\boxed{\Leftrightarrow}$. Whatever menu you have will disappear, and the top six keys on the right will now control the cursor: the keys marked (above in white) $\triangle, \triangledown, \triangleleft, \triangleright$ move the blinking cursor up, down, left, right. The key marked DEL deletes whatever character the cursor is blinking on, and the key marked INS toggles back and forth between "insert" and "overwrite" mode (insert mode means each character typed is inserted in the position that the blinking arrow cursor points at). Later when we write and edit programs, we will use this cursor control frequently. Right now, backspace is probably all you need. Please note, however, that all of this you can only do to the *command line*; you cannot move the cursor through the *stack* or overwrite or insert or delete items on the *stack*. The command EDIT (red function of ENTER) takes level 1 of the stack and puts it on the command line, leaving behind a copy on the stack. This is the only way you can edit the stack, and it only works on level 1. It is especially useful to view something, like a matrix, which is at level 1 of the stack but too big to display on the screen.

Oh yes, if you are tired of the HP-28S beeping every time you make a mistake, press $\boxed{51}$ $\boxed{\text{ENTER}}$ $\boxed{\text{S}}$ $\boxed{\text{F}}$ $\boxed{\text{ENTER}}$. This turns off the beeper; error messages will still be displayed.

Finally, if you are having trouble, you can always consult the *Owner's Manual*. The most important section is the answers to common questions on pages 282-286. READ IT!

### 8.1.2  Direct Entry Mode for Arithmetic and Algebraic Expression:

The HP-28S can also deal with arithmetic expressions, such as $'2*3 - 6/5'$ or $'SIN(\pi/6)'$. To put an arithmetic expression on the stack, press the single quote key, $\boxed{'}$, just below the ENTER key, followed by the expression written in the usual algebraic, rather than reverse Polish order. Terminate the expression with another single quote and press ENTER to put the expression on the stack. Actually, you do not have to provide the terminating single quote before pressing ENTER; the HP-28S *always attempts to fill in terminating brackets* on the command line when ENTER is pressed (other kinds of brackets are (, [, {, $\ll$, and "—each is associated with a different data structure). Thus, $\boxed{'}$ $\boxed{2}$ $\boxed{\times}$ $\boxed{3}$ $\boxed{\text{ENTER}}$ puts $'2*3'$ on the stack.

Arithmetic expressions can be operated on just like numbers. Thus if $'3/5'$ is at level 2 of the stack, and $'5/3'$ at level 1 and you press $\boxed{\times}$, you will see $'3/5*(5/3)'$. Notice the extra parenthesis and the symbol $*$ for the key $\boxed{\times}$. Also notice that no simplifications are performed.

To find the numerical value of an expression press $\boxed{\text{EVAL}}$. Thus if $'3+5'$ is at level 1 of the stack and $\boxed{\text{EVAL}}$ is pressed you will see 8 (or maybe 8.000 if you're in the 3 FIX mode). You can even use EVAL on unfinished expressions on the command line, since EVAL, like ENTER, provides terminating brackets. Thus if you press $\boxed{'}$ $\boxed{2}$ $\boxed{\times}$ $\boxed{(}$ $\boxed{5}$ $\boxed{+}$ $\boxed{8}$ $\boxed{\text{EVAL}}$ you'll get 26. In particular, if you are not fond of RPN, you can do computations by pressing $\boxed{'}$ first, then typing in numbers and operations in the order you normally use on other calculators, and then pressing $\boxed{\text{EVAL}}$ just the way you would press $\boxed{=}$ on other calculators. The only extra expense is the initial single quote.

You may have noticed that after you begin an expression with $\boxed{'}$, the blinking cursor on the command line changes design. This is to indicate that you are now in "algebraic" entry mode, instead of "immediate" entry mode. Usually when you press $\boxed{+}$, the calculator tries to add two things. Inside the single quote, when you press $\boxed{+}$, the symbol $+$ appears and the operation $+$ is disabled. As soon as you press ENTER, EVAL, or a terminating single quote, the calculator reverts to the mode you are used to.

Expressions can also contain variables, for example $'3+5*X'$ or $'SIN(T/2)'$. (The letters X and T are the usual letters on the left keyboard). It is here that the HP-28S begins to part company with standard scientific calculators. These algebraic expressions can be operated on just like other arithmetic expressions. The X is actually a variable, and values can be stored as X. If you press $\boxed{3}$ $\boxed{\text{ENTER}}$ $\boxed{'}$ $\boxed{X}$ $\boxed{\text{STO}}$, you will store the number 3 in a memory location named X. The name X will now appear on your USER menu; if you press $\boxed{\text{USER}}$, you will see a menu with the item X at the far left. If you press the button under that menu item, you will get the present value of X, which is 3. Notice in the HP-28S, that mentioning the name of variable without single quotes, puts the *value* of that variable on the stack. To see this, type $\boxed{X}$ from the left keyboard, then $\boxed{\text{ENTER}}$. To refer to the *name* of X, rather than its value, you must use $'X'$.

We will discuss the USER menu later. It is where you store programs as well as variables, and it can be organized into a tree of directories and subdirectories. The one thing you may wish to know now is how to clear a variable, such as X, out of your USER area. Simply press $\boxed{'}$ $\boxed{X}$ $\boxed{\text{PURGE}}$ (red function of $\boxed{4}$).

If the variable X in the expression $'3+5*X'$ has a value, say 2, then the expression also has a numerical value. To find that number, just press EVAL as usual. If the contents of X is another variable Y and Y has value 7, then EVAL only takes you one step so that you get $'3+5*Y'$. If you now press EVAL again, you get 38. If you wish to evaluate all the way, press $\blacksquare$ $\boxed{\rightarrow\text{NUM}}$. This forces a numerical value, whenever possible, as you saw in the last section with $\pi$.

The HP-28S has an ALGEBRA menu (red function of $\boxed{J}$), which contains various commands for forming, expanding, and collecting terms. We will not discuss these commands since the HP-28S is a rather clumsy algebraic manipulator—just try expanding $'(X+1)\wedge 5'$.

One important thing the HP-28S can do with expressions is to set them equal to zero and solve them. There is an entire menu for this, the SOLV menu, but right now we are interested in just one command on this menu, ROOT. To solve $\sin(X) = 0$ for the root nearest to 6, first press $\boxed{'}$ $\boxed{X}$ $\boxed{\text{ENTER}}$ $\boxed{\text{TRIG}}$ $\boxed{\text{SIN}}$ (or $\boxed{'}$ $\boxed{S}$ $\boxed{I}$ $\boxed{N}$ $\boxed{(}$ $\boxed{X}$ $\boxed{\text{ENTER}}$). Then press $'X$ followed by ENTER. Then press 6 followed by ENTER. Now press $\boxed{\text{SOLV}}$ $\boxed{\text{NEXT}}$ $\boxed{\text{ROOT}}$, or if you prefer, type $\boxed{R}$ $\boxed{O}$ $\boxed{O}$ $\boxed{T}$ on the alphabetic keyboard followed by $\boxed{\text{ENTER}}$. After a few seconds, you should see your answer, which in this case is $2\pi$ to 12 digits.

To repeat, the syntax for the ROOT command is, at stack level 3, the expression being set equal to zero, at level 2, the variable being solved for (inside single quotes), and at level 1, a number close to the desired root. Note that if you are solving for, say, X, then the ROOT command will put the variable X on your USER menu, so after you finish the ROOT command you may want to PURGE the variable X.

## 8.2  Built-in Graphing

In Section 8.3, we will show you how to customize your HP-28S to make it as convenient for graphing as other graphing calculators, as well as a great deal more powerful for numerical computations. Nevertheless, the built-in graphing capability of the HP-28S is not so shabby. Let's look at the PLOT menu.

Suppose we wish to graph $y = \sin x$. First type $\boxed{'}$ $\boxed{X}$ $\boxed{\text{ENTER}}$ $\boxed{\text{TRIG}}$ $\boxed{\text{SIN}}$. Then press $\blacksquare$ $\boxed{\text{PLOT}}$ (the red function of $\boxed{\text{SOLV}}$). Now press $\boxed{\text{STEQ}}$ (for "store equation") on the PLOT menu. If you check your USER menu, you will now see the item EQ. If you press $\boxed{\text{EQ}}$ on your USER menu, you will see the expression stored in EQ, namely $'\text{SIN}(X)'$. Go back to the PLOT menu and press $\boxed{\text{DRAW}}$. The stack display will disappear, axes will be drawn, and $y = \sin x$ will be plotted. Make sure you are in radians mode or else the plot will look like $y = 0$. After the calculator is finished drawing the graph, try pressing the cursor control keys (just below the screen). You should see a cross-shaped cursor, called "cross-hairs," move around the screen. Move it to the top of the sine wave at $x = \pi/2$. Then press $\boxed{\leftrightarrow}$ and hold it down. You will see on the bottom of the screen the ordered pair of numbers for the point where the cross-hairs are located (it should be about $(1.6, 1)$). If you press $\boxed{\text{INS}}$, this ordered pair will be placed on the stack; you won't see this until you return to the stack display.

**Zooming-in.** Return to the stack by pressing $\boxed{\text{ON}}$. The point with coordinates of approximately $(1.6, 1)$ should be on the stack. Suppose we want to zoom in on the graph around this point. First press $\boxed{\text{CENTR}}$ on the PLOT menu (you'll need to use $\boxed{\text{NEXT}}$ since it is on the second row of the menu). This centers the plot around whatever point is at level 1 of the stack, in this case (1.6, 1). Now press $\boxed{0.5}$ followed by $\boxed{* \text{W}}$ on the same part of the PLOT menu. This cuts down the width of the viewing rectangle by the factor 0.5. Pressing $\boxed{0.2}$ followed by $\boxed{* \text{H}}$ cuts down the height of the viewing rectangle by the factor 0.2. Then press $\boxed{\text{DRAW}}$ again. You will see one hump of the sine curve, and the cross-hairs cursor. You could now recenter and zoom-in again, repeating this process as often as you like.

There is another way of cutting down the viewing rectangle. With one hump of the sine curve still on the screen, imagine a rectangle containing a small piece of the curve you wish to focus on. Move the cross-hairs to the *lower left hand corner* of this imaginary rectangle and press the $\boxed{\text{INS}}$ cursor key. Then move the cross-hairs to the *upper right* of the imaginary rectangle and press $\boxed{\text{INS}}$ again. Then press $\boxed{\text{ON}}$ to return to the stack. You will see two ordered pairs on the stack. The one at level 1 corresponds to the location of the cross-hairs the last time you pressed $\boxed{\text{INS}}$, namely the upper right corner of the desired viewing rectangle. To reset the viewing rectangle's right corner to that point, press $\boxed{\text{PMAX}}$ on row 1 of the PLOT menu. The ordered pair now remaining on the stack can be made into the lower left corner of the HP-28 viewing rectangle by pressing $\boxed{\text{PMIN}}$. If you now press $\boxed{\text{DRAW}}$, the graph will be redrawn in the new viewing rectangle.

At this point, you may wonder how the calculator set up the original viewing rectangle. Return to stack display and pull up your USER menu by pressing $\boxed{\text{ON}}$ $\boxed{\text{USER}}$. In addition to the variable EQ we saw before, you will also see the variable PPAR (plot parameters). Press the key under PPAR. You will see a list of items contained in a pair of curly brackets. These are your plot parameters. The first is your PMIN point; the second is your PMAX point (each coordinate of the PMIN point must be less than the corresponding coordinate of the PMAX point or else you'll get an error message when you DRAW). The remaining items are not usually of interest (independent variable, how many pixels to plot where 5 means every 5th pixel, where to center the axes, which is nearly always (0,0)).

When you first pressed DRAW, there were no plot parameters so the calculator provided default values. To see what they are, let's get rid of your present PPAR; press $\boxed{'}$ $\boxed{\text{PPAR}}$ $\boxed{\blacksquare}$ $\boxed{\text{PURGE}}$. Now press $\boxed{\text{DRAW}}$ on the PLOT menu. After the plot is finished, or even while the calculator is still plotting, press $\boxed{\text{ON}}$ and bring back up your USER menu. Press $\boxed{\text{PPAR}}$ and you will now see the default plot parameters. You are mostly interested in the PMIN and PMAX points. You should see (-6.8, -1.5) and (6.8, 1.6), which corresponds to the viewing rectangle $[-6.8, 6.8]$ by $[-1.5, 1.6]$. Here is the reason why these are the default values: The HP-28S screen is 137 pixels wide and 32 pixels high. With the default values, each pixel is 0.1 in width and height. The tick marks on the horizontal and vertical axes are located every 10 pixels, so for the default PPAR the tick marks are at consecutive integers.

Most of the other options on the PLOT menu are not needed here. Note that PPAR is on the PLOT menu as well as your USER menu and that you can see what is in EQ by pressing RCEQ (recall equation) on the PLOT menu instead of pressing EQ on the USER menu. All of the commands involving $\Sigma$ are for scatterplots of ordered pairs stored in the statistical matrix $\Sigma$ DAT. The command PIXEL is very important for polar coordinate or parametric equation plots; it takes an ordered pair as input and blackens the pixel on the graphics screen located at that point. Putting the command PIXEL in a loop that increments an independent variable is how you plot, point by point, a parametrically defined curve. The command CLLCD clears the graphic screen in preparation for such a custom plot and DRAX draws the axes.

It is possible to graph two functions simultaneously. For example, if you wish to graph both $y = \sin x$ and $y = \cos x$ on the same screen, enter $'\text{SIN(X)} = \text{COS(X)}'$ followed by $\boxed{\text{ENTER}}$ (or enter $'\text{SIN(X)}'$ and $'\text{COS(X)}'$ separately followed by $\boxed{=}$ $\boxed{\text{ENTER}}$). Note that the $\boxed{=}$ key is on the left side, and has nothing to do with the $\boxed{=}$ key you see on most scientific calculators. Now press $\boxed{\text{STEQ}}$ and then $\boxed{\text{DRAW}}$, and you will see sin x and cos x graphed simultaneously. Many other commands, such as ROOT, operate on "equations" as well as expressions.

## 8.3 A Customized Graphing Directory

The power of the HP-28S is the flexibility that comes with its programming capability and its memory management. In this section, we will show you how to use these functions to create a convenient environment for graphing.

**Establishing the MAIN Directory.** The USER memory can be organized into a tree of directories and subdirectories using the command CRDIR. There is one "top" directory, which is automatically called HOME. All programs or variables in a given directory are accessible to all its subdirectories but not vice versa. For this reason, we suggest you set up your USER memory as follows (it is best to do it now, because it is much harder to do later when your calculator is full of programs). First, press $\boxed{\text{USER}}$. Then type ' MAIN ' followed by $\boxed{\text{CRDIR}}$ (on the MEMORY menu, left keyboard, red function of $\boxed{\text{I}}$). This creates a subdirectory of the HOME directory called MAIN. From now on, you will spend most of your time in the MAIN directory. The rest of the HOME directory will contain general utility programs, which, because they are located in HOME, will be available to all your directories.

**A Critical Program.** Here is an essential utility. It quits whatever directory you are presently in and returns to the directory immediately above where you are. Its name is QT.

$$\ll \ \{\text{HOME}\} \ \text{PATH} + \text{DUP SIZE 1} - \text{GET EVAL} \gg$$

To store this program, first type it in as written. The minute you type the left programming bracket $\boxed{\ll}$ you enter a different entry mode somewhat like the algebraic entry mode inside single quotes. One difference you will see is when you press $\boxed{+}$ : spaces are automatically provided on either side of the + symbol. You need not provide the terminating bracket $\gg$ before pressing $\boxed{\text{ENTER}}$, but you do need to provide the terminating bracket } around HOME. After you have entered the program on the stack type $\boxed{'}\boxed{\text{Q}}\boxed{\text{T}}\boxed{\text{STO}}$. You should now see the name QT on your USER menu.

**Using the VISIT Feature.** If you want to view or change a program you have written, such as QT, type $\boxed{'}\boxed{\text{Q}}\boxed{\text{T}}\boxed{\text{VISIT}}$. The program will then appear on the command line and the cursor control keys will automatically be activated so that you can edit the program, just as if you had pressed $\boxed{\leftrightarrow}$. When you are finished, press $\boxed{\text{ON}}$ if you want the program to remain as it was, or press $\boxed{\text{ENTER}}$ if you want the version now on your command line to replace the old version.

**The GRAPH Directory.** We are now ready to create our graphing directory. First press $\boxed{\text{MAIN}}$ on your USER memory. You will descend to the MAIN subdirectory of the HOME directory; the subdirectory menu is empty of course. Now type $\boxed{'}$ $\boxed{\text{G}}$ $\boxed{\text{R}}$ $\boxed{\text{A}}$ $\boxed{\text{P}}$ $\boxed{\text{H}}$ followed by pressing $\boxed{\text{CRDIR}}$ on the MEMORY menu. Then press the new $\boxed{\text{GRAPH}}$ key on the USER memory to descend to the GRAPH subdirectory. Just for fun, type $\boxed{\text{Q}}$ $\boxed{\text{T}}$ $\boxed{\text{ENTER}}$, and watch yourself return to the MAIN menu. If you type $\boxed{\text{Q}}$ $\boxed{\text{T}}$ $\boxed{\text{ENTER}}$ again you'll return to the HOME menu.

Take a moment to consider the current status of your USER memory. It now contains a top directory, called the HOME directory, as well as two subdirectories, MAIN and GRAPH. The HOME directory contains at least four different types of objects: QT, a program; MAIN, a directory; PPAR, a list; and EQ, an algebraic object. The MAIN directory is a subdirectory of the HOME directory. MAIN currently contains only one subdirectory, GRAPH, which is empty. Later, as need or desire arises, you may wish to add other subdirectories to the MAIN directory in addition to GRAPH. GRAPH and the other subdirectories will consist primarily of sets of programs, but once used, will likely contain other types of objects as well— variables, equations, etc. It is wise to plan the organization of your USER memory carefully.

Now return to the (empty) GRAPH directory. The 10 programs listed and explained below all should be entered in this directory. If you store any of these programs while in a different directory, they will be stored there instead of in GRAPH, where they belong.

To store each program, enter the complete code on the command line, then press ENTER to place the program code on level 1 of the stack. Then press ⎡'⎤ followed by the program name (e.g., CLDRA) followed by STO to store the program. If the GRAPH directory menu is showing on your screen, you will see the new program name added to the menu as you store it.

If you have made a syntax error in the program code, the HP28S will give you an opportunity to correct it when you try to enter it onto the stack. Errors in spacing are the most common syntax errors; they can be corrected by inserting or deleting spaces as needed. If you have made some other type of error in a program, you can correct it by using the VISIT command after the program has been stored. Editing programs by using the VISIT command was explained above.

**CLDRA** (Clear and Draw). Our first program simply stores a new expression to graph, graphs it, and then stores a string of data encoding the LCD screen of the graph. Its only input is the expression to be graphed.

<< DUP STEQ 1 → LIST 'EQS' STO CLLCD DRAW LCD → 'SCR' STO DGTIZ >>

The command → **LIST** is on the LIST menu. If you try to type it in letter-by-letter, the program entry mode automatically provides spaces around the arrow, which you do not want. If you press the cursor control key, you can move the cursor back and delete the space between → and LIST. If instead you press ⎡→LIST⎤ on the LIST menu, don't forget that you now lose cursor control. It is easy to forget this and press menu buttons expecting to move the cursor and instead fill your program with garbage commands. Press ⎡◀▶⎤ to return to the cursor control keys. Generally, it is easier to type than to pull up menus, except when the command contains a symbol which program entry mode wants to surround by spaces. The command **LCD** → on the STRING menu is another example of such a command.

**RCLGR** (Recall Graph). Our next program acts like a toggle switch to display the present graph without having to redraw it. It has no input.

<< SCR → LCD DGTIZ >>

By the way, the command DGTIZ is necessary to activate the cross-hairs.

**DFPAR** (Default Parameters). The following program resets the plotting parameters to the default values. It has no input.

<< {(−6.8, −1.5) (6.8, 1.6)   X   1   (0,0)} 'PPAR' STO >>

To type the minus signs use the change sign key CHS (row 3, right side).

**OVDRA** (Overdraw). This program plots the graph of a new function over the present graph. The new expression is placed at the beginning of the list EQS, which contains all the expressions plotted so far on the same screen. The only input for OVDRA is the new expression.

$$<< \text{ DUP STEQ } 1 \rightarrow \text{LIST EQS} + \text{ 'EQS' STO SCR } \rightarrow \text{LCD DRAW}$$

$$\text{LCD} \rightarrow \text{ 'SCR' STO DGTIZ } >>$$

**REDRA** (Redraw). This program redraws the graphs of all the function in the list EQS using whatever the present plot parameters are. It is used after the plot parameters have been changed. It has no input.

$$<< \text{ CLLCD EQS LIST} \rightarrow 1 \text{ SWAP START STEQ DRAW NEXT}$$

$$\text{LCD} \rightarrow \text{ 'SCR' STO DGTIZ } >>$$

The START...NEXT command is the simplest of various loop structures built into the HP-28S. The command LIST $\rightarrow$ pulls a list apart and puts the individual items on the stack with the number of items in the list at level 1 of the stack.

**BOX.** This program takes any two points (ordered pairs of numbers) and resets the plotting parameters so that those two points form diagonally opposite corners of the plotting rectangle. The program then draws the graphs of all functions in the list EQS. The input is two ordered pairs in any order.

$$<< \rightarrow \text{P} \quad \text{Q} << \text{ P RE Q RE MIN P IM Q IM MIN R} \rightarrow \text{C PMIN P}$$

$$\text{RE Q RE MAX P IM Q IM MAX R} \rightarrow \text{C PMAX REDRA } >> \quad >>$$

Within the program, the command $R \rightarrow C$ occurs twice; there should be no space around the arrow. This command is on the COMPLX menu. It takes two real numbers and forms an ordered pair. In this case the ordered pairs formed by $R \rightarrow C$ are used as PMIN and PMAX to set the new viewing rectangle.

**ZOOM.** This program takes a point (ordered pair) at stack level 3, an X-magnifying factor at level 2, and a Y-magnifying factor at level 1 and zooms into the point using the given magnifications. All functions in EQS are redrawn.

$$<< 3 \text{ ROLL CENTR INV *H INV *W REDRA } >>$$

Note the the factor 2 means "zoom in" while the factor .5 means "zoom out".

**NEWX:.** This program resets the X-limits of the plotting rectangle. Input is two numbers in any order.

$$<< \rightarrow \text{NUM SWAP } \rightarrow \text{NUM DUP2 IF} > \text{THEN SWAP END PPAR 2}$$

$$\text{GET IM R} \rightarrow \text{C PMAX PPAR 1 GET IM R} \rightarrow \text{C PMIN } >>$$

**NEWY.** This program resets the Y-limits of the plotting rectangle. Input is two numbers in any order.

$$<< \rightarrow \text{NUM SWAP } \rightarrow \text{NUM DUP2 IF} > \text{THEN SWAP END PPAR 2}$$

$$\text{GET RE SWAP R} \rightarrow \text{C PMAX PPAR 1 GET RE SWAP R} \rightarrow \text{C PMIN } >>$$

**Y.0.** This program takes a point or number near a root of the expression in EQ and refines it to 12 digit accuracy. Its input is an ordered pair or number. Its output is the root.

$$\ll \text{RE} \rightarrow \text{NUM EQ 'X' 3 ROLL ROOT 'X' PURGE} \gg$$

The rest of the GRAPH directory will contain variables for storing information for plotting: EQ, EQS, PPAR, SCR. When you are finished entering all of these programs they will not be in the order you would like on the menu. To get them in the order you want, press $\boxed{\{}$ $\boxed{\alpha}$ followed by the menu keys in the order you want them. When you are finished press $\boxed{\text{ENTER}}$. A list of menu items is now on level 1 on the stack. If you now press $\boxed{\text{ORDER}}$ on the MEMORY menu, the GRAPH directory will be reordered according to the list at level 1 of the stack. You reorder any USER directory in a similar manner. You need not put all items on your directory menu in the list. Any omitted items are automatically put at the end of the menu, in the same order relative to each other as they presently occupy.

# Chapter 9

## Graphing with the HP 48 Scientific Calculator

This chapter is an introduction to the use of the HP 48S and HP 48SX calculators for graphing. The two versions of the HP 48 calculator differ only in that the HP 48SX offers plug-in memory card slots which you may use to add more memory or pre-programmed functionality. For our purposes here, the two operate identically. We will refer to them collectively as the HP 48.

Unlike most of the other calculators described in this manual, the HP 48 is not designed primarily as a graphing tool. Instead, it is intended to be a general purpose mathematics calculator, incorporating extensive graphics capability as just one aspect of its overall computing resources. This generality means that the HP 48 requires a larger learning-time investment than other calculators. But that investment pays off in the application of the HP 48 to a broad range of theoretical and applied problems that goes well beyond introductory mathematics courses.

It is very difficult to compress an adequate description of HP 48 fundamentals and graphing techniques into this one chapter. The calculator comes with a comprehensive two-volume manual and a quick-reference guide. You are encouraged to read at least Chapters 1 through 6 of the Owner's Manual for a more thorough introduction to HP 48 operation. Additional references are given at the end of this chapter.

### 9.1  HP 48 Fundamentals

**9.1.1  The HP 48 Keyboard:** Because the HP 48 contains so many computing functions, its keyboard provides multiple uses for each of the 49 keys.

(1) **Primary functions.** Except for the top row of 6 keys, all keys have their primary function labeled on the key top. If you press a key without prefixing it with any of the shift keys, it executes the labeled function.

(2) **Menu keys.** The blank keys in the top row are called *menu keys*. Their actions are determined by the *menu labels* shown at the bottom of the display screen. If you press the (unshifted) [MTH] key, you will see these labels:

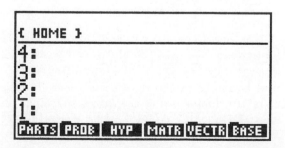

A little "tab" at the top of a menu label indicates that pressing the menu key below it activates yet another menu of functions. For example, pressing the menu key labeled [HYP] activates a menu of hyperbolic functions:

```
┌─────────────────────────────────┐
│ { HOME }                        │
│ ─────────────────────────────── │
│ 4:                              │
│ 3:                              │
│ 2:                              │
│ 1:                              │
│ SINH ASINH COSH ACOSH TANH ATAN │
└─────────────────────────────────┘
```

Now the leftmost menu key executes SINH, the next ASINH(arc-sinh), etc.

Some menus contain more than one "page" of six keys. You can use the [NXT] key to step through the contents of any menu.

- **Shifted Keys.** Two "shift keys" [←] and [→], placed in the lower left corner of the keyboard and colored orange and blue, respectively, modify the actions of the other keys. The additional actions are labeled in the corresponding color above each key. For example, pressing [NXT] advances the current menu to the next set of six menu labels; pressing [←] first then [NXT] goes back to the previous set. In this chapter we will refer to the shift combinations by the colored label rather than showing the primary key label, e.g. [←][PREV] rather than [←][NXT], to better indicate the intended function.

  Several keys have blue-shifted functions even though they are not explicitly labeled. For example, [→][PLOT] activates the "plot execution" menu for graphing functions ([←][PLOT] is for the "plot equation entry" menu). The unlabeled functions are usually variations on the orange-labeled functions, and are shortcuts for alternate methods. The plot execution menu is available as the [PLOTR] menu key in the plot equation entry menu.

  To cancel an unwanted shift key press, just press the shift key again.

- **Alpha Keys.** Most keys, in addition to their primary and shifted functions, may be used to enter alphabet letters and other symbols. This is achieved by means of the [α] key, which acts in effect as a third shift key. Pressing [α] followed by another key enters the character printed in white at the lower-right corner of the key. Thus, pressing [α] followed by [cos] enters a "T" character. To enter a lower-case character, press [←] between [α] and the character key; for example, to enter an "x," press [α] then [←] then [1/x].

  To type a series of consecutive characters, you can hold down [α] while you press the character keys. Alternatively, you can press [α] twice; this locks the alpha shift so that all subsequent key presses enter alpha characters. Alpha lock is cleared by [ENTER] or by [α] pressed again.

  To improve the legibility of keystroke examples, we will generally represent the entry of letters and digits as unadorned characters rather than showing key boxes and alpha shifts. For example, "123 ABC ENTER" indicates that you are to press [1][2][3][SPC][α][α][A][B][C][ENTER].

- **Cursor keys.** The keys marked ◁ ▷ △ and ▽ are for moving a *cursor* in the directions indicated by the triangles on the keys. The cursor takes various forms: in the command line (see below), it is a blinking arrow that shows where character insertion takes place; on the graphing screen, the cursor is a small cross +; in the equation catalog, it is a triangle that points at one entry.

  Pressing ↱ followed by any cursor key moves the cursor as far as possible in the indicated direction.

- The ON key not only turns the HP 48 on, but acts as a general purpose interruption key (note the ATTN—for *Attention!*—marked below the key). It halts calculator execution and returns to a previous state. For example, when you are entering a number, pressing ON clears the number and returns to the normal stack display. Anytime the calculator is in a state that you wish to abandon, press ON one or more times, and it will return to its default state showing the stack.

**9.1.2  The Stack:** If you have not previously used a Hewlett-Packard calculator, you may find ordinary arithmetic to be a little awkward at first on the HP 48. The HP 48 uses a uniform approach to calculating in which you apply *functions*, i.e. $+$, $-$, SIN, etc., to arguments that you have previously entered. Prior to function execution, the arguments are held on a "stack," which grows and shrinks as you add or remove entries. For example, starting with the display like this,

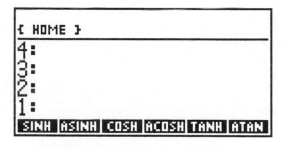

typing     1     ENTER     2     ENTER     3     ENTER

enters the three numbers onto the stack:

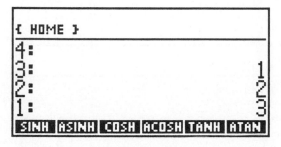

Now if you press +, the sum of the bottom two numbers replaces those numbers:

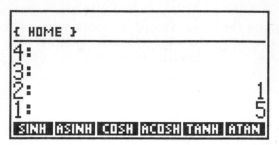

The logic resembles that of ordinary pencil-and-paper arithmetic—for example, to add two numbers, you write them down first then perform the addition. The advantages are several:

(1)  A wide variety of operations, mathematical and non-mathematical, can be incorporated into the same scheme.

(2)  You can work your way through a calculation without knowing all of the steps in advance.

(3)  The result of each calculation is returned to the stack, where it is immediately available for further operations. You never have to write down or re-enter intermediate results.

The last point is easily illustrated: press ⊞ again:

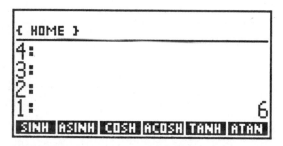

The first sum (5) is added to the value that had been entered previously.

There is no limit other than available memory on the number of items that you can enter into the stack. When there are more than four, you can see the fifth and higher by pressing △ repeatedly. To remove the level 1 item from the stack, press ⬅ when no command line is present. To clear the entire stack, press ➡ CLR.

The HP 48 stack can hold a variety of data types as well as ordinary real numbers. For example, the complex number *1+2i* is represented as the ordered pair (1,2): enter ⬆ ( ) 1 SPC 2 ENTER :

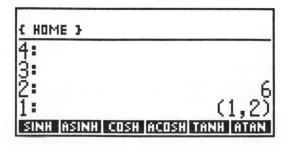

The logic for dealing with complex numbers is the same as for real numbers: enter the arguments then execute the functions: e.g. with the (1,2) still in level 1, (4 5) ⊞ returns the complex sum:

```
┌──────────────────────────────┐
│ RAD                          │
│ { HOME }                     │
│──────────────────────────────│
│ 4:                           │
│ 3:                           │
│ 2:                        6  │
│ 1:               (5,7)       │
│ SINH ASINH COSH ACOSH TANH ATAN │
└──────────────────────────────┘
```

**9.1.3  The Command Line:** When you type passive data like the digits in a number, the characters appear below the first stack level, in an editing field called the *command line*. A blinking cursor appears, to indicate where characters will be inserted. Initially, the cursor is at the end of the line, but you can move it back or forwards with ◁ and ▷ to correct an entry. The backspace key ⇐ erases the character to the left of the cursor; DEL erases the character under the cursor.

To enter a negative number, type the number's digits, then press +/-.

You can enter functions as well as data into the command line. For example, typing SIN then pressing ENTER computes the sine of a number in level 1, just as if you had pressed SIN. This is handy when you can't remember which menu contains a particular function—just type its name and press ENTER.

To cancel a command line entry, press ON.

**9.1.4  Entering Expressions:** Although you can make arbitrarily complicated calculations using only stack operations, it is often convenient to enter a calculation as a formula. The HP 48 provides for this: you can enter any expression or equation by pressing the single-quote key ', then typing the formula between the two quotes. For example, to compute

$$\frac{3+5^2}{1+6},$$

enter ' ⇦ () 3+5 $y^x$ 2 ▷ ÷ ⇦ () 1+6 ENTER. This enters the expression in unevaluated form:

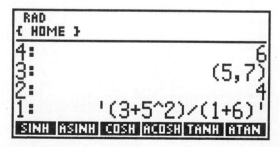

```
┌──────────────────────────────┐
│ RAD                          │
│ { HOME }                     │
│──────────────────────────────│
│ 4:                        6  │
│ 3:               (5,7)       │
│ 2:                        4  │
│ 1:         '(3+5^2)/(1+6)'   │
│ SINH ASINH COSH ACOSH TANH ATAN │
└──────────────────────────────┘
```

To compute the value, press EVAL, which returns 4. You can skip the intermediate stage by pressing EVAL instead of ENTER. Usually, when you enter an expression without evaluation, it is to store it as a function for further analysis such as graphing.

You can also enter an expression in a manner resembling common written form by using the Equation-Writer. In this system, you "draw" an expression in two dimensions. Most of the drawing is self-evident, but you do need to know the following important operations:

(1) Press $\boxed{\leftarrow}$ $\boxed{\text{EQUATION}}$ to activate the EquationWriter.

(2) To start entry of an exponent, press $\boxed{y^x}$.

(3) To start entry of a fraction, press $\boxed{\triangle}$, then proceed to enter the numerator.

(4) To complete any "field"—finish an exponent, end a denominator, etc.—to move on to the next part of an expression, press $\boxed{\triangleright}$.

(5) You can correct a previously typed character by pressing $\boxed{\leftarrow}$. However, backspacing over a function can be quite slow, since the entire picture has to be rebuilt as if you had retyped in from the start.

(6) To complete entry and return to the stack display, press $\boxed{\text{ENTER}}$. To cancel the current entry, press $\boxed{\text{ON}}$.

(7) To view more of an entry that has grown bigger than the screen, press $\boxed{\leftarrow}$ $\boxed{\text{GRAPH}}$, then use the arrow keys to scroll the display window. To return to expression entry, press $\boxed{\leftarrow}$ $\boxed{\text{GRAPH}}$ again.

For example, the previous example may be entered with $\boxed{\leftarrow}$ $\boxed{\text{EQUATION}}$ $\boxed{\triangle}$ 3+5 $\boxed{y^x}$ 2 $\boxed{\triangleright}$ $\boxed{\triangleright}$ 1+6 $\boxed{\triangleright}$:

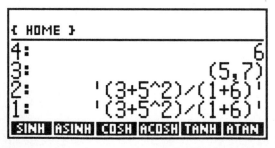

Then $\boxed{\text{ENTER}}$ converts the picture to a stack entry:

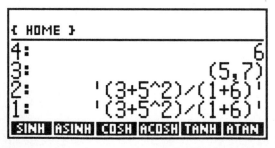

As in the case of command line entry, you can evaluate an EquationWriter expression directly by pressing $\boxed{\text{EVAL}}$ instead of $\boxed{\text{ENTER}}$.

**9.1.5  Variables:** Variables in the HP 48 are represented by alphanumeric names, which may have up to 127 characters, but must start with a non-numeral character, usually a letter. Names also may not contain any of the data punctuation characters, including ', ", [, ], (, ), # , { , } , :, << and >> . They must not be the same as the name of any built-in functions, such as SIN or RE. Typical variable names are X, Y, T0, and Temperature. Names may appear within expressions, such as $'X+Y-Z0'$. There is no limit on the number of variables that you can define.

If you enter a name in the command line and press ENTER, the HP 48 returns to the stack the value currently associated with that variable. However, there are many circumstances in which you want the name itself on the stack instead of the value. To do this, enter the name surrounded by single quotes. One such case is assigning a value to a variable, which is achieved with STO (*store*):

$$123' \text{ ABC}' \text{ STO}$$

assigns the value 123 to the variable ABC. If you then enter the unquoted name ABC ENTER, the HP 48 returns the value 123 to the stack.

Anything that you can put on the stack can be stored in a variable. For example, $'1+2*X^3'$ $'$CUBIC$'$ STO stores the expression $'1+2*X^3'$ in a variable named CUBIC.

When you assign a value to a variable using STO, the name and the value are stored in memory, and the name appears in a menu activated by the VAR key. For example, after storing variables ABC and CUBIC, VAR shows this menu:

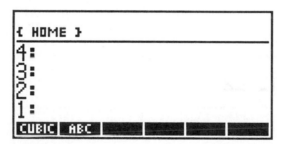

The VAR menu keys provide shortcuts for entering the names:

(1)  Pressing an unshifted VAR menu key returns the stored value as if you had typed the labeled name in the command line and pressed ENTER.

(2)  Pressing ▶ followed by a menu key also returns the value. (The difference between the unshifted and right-shifted menu keys is only relevant for programs and stored names, which we are not discussing here.)

(3)  Pressing ◀ followed by a menu key replaces the current value of the corresponding variable with the item in level 1 of the stack.

(4)  Pressing ' followed by a menu key enters the quoted name into the command line.

The preceding sections have barely scratched the surface of HP 48 methods and capabilities. The material, however, should provide a sufficient introduction so that you can proceed with the main topic, HP 48 graphing.

## 9.2  Graphing Essentials

When you set out to make a graph on the HP 48, as on any other graphing system, you must specify a number of elements that determine the nature and appearance of the graph. Obviously, you must provide either a set of coordinates for points to be graphed, or a mathematical expression from which the coordinates can be computed. But, in addition, you must determine:

(1)   The manner in which the coordinates or expression is to be interpreted. For example, does an expression represent $y$ (vertical coordinate) as a function of $x$ (horizontal coordinate), or $r$ (radial coordinate) as a function of *theta* (polar angle)?

(2)   Which variables in the expression correspond to the two coordinate axes. Although it is common to use $x$ and $y$ as the variables corresponding to the horizonal and vertical coordinates, there are many cases when the use of other variables is desirable.

(3)   The plot scale, i.e. the mapping of the logical coordinates associated with a graph onto the physical display.

(4)   Whether axes are to be drawn, and if so, where.

To illustrate, let's try a sample graph of sin $x$ . First, enter the expression:

`'` `SIN` `α` `X` `ENTER` `←` `PLOT` `STEQ`

These keystrokes create the expression $'SIN(X)'$ and store it in the variable EQ, which is special to the graphing system. Next, make sure that the HP 48 will produce an ordinary function graph:

`PTYPE` `FUNC`

At this point, you should see a display like this:

```
Plot type: FUNCTION
EQ: 'SIN(X)'
4:
3:
2:
1:
PLOTR PTYPE NEW EDEQ STEQ CAT
```

Now enter the following:

`PLOTR` `NXT` `RESET` `←` `PREV`

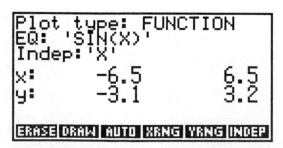

```
Plot type: FUNCTION
EQ: 'SIN(X)'
Indep: 'X'
x:        -6.5              6.5
y:        -3.1              3.2
ERASE DRAW AUTO XRNG YRNG INDEP
```

You have reset the calculator to its default plotting ranges.

The last step before graphing is to select *radians* angle mode to indicate that X is to be interpreted in radians. Press $\boxed{\text{ON}}$. If you do not see the symbol RAD in small letters at the top left part of the display, then press $\boxed{\leftarrow}$ $\boxed{\text{RAD}}$ to turn it on.

Now you are ready to draw a graph. Press $\boxed{\text{ERASE}}$ $\boxed{\text{DRAW}}$:

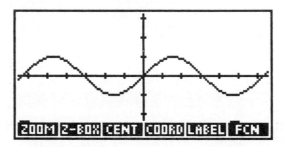

After drawing the graph, the HP 48 activates *interactive graph mode* (see Section 9.3), which allows you to rescale the graph (zoom), read data from the graph, and compute intercepts, slopes, and areas. To return to the normal calculator mode, press $\boxed{\text{ON}}$.

In the next sections, we will study in more detail each of the graphing steps illustrated in the preceding example. We will first use ordinary function graphs to illustrate general methods, then consider other types of graphs.

### 9.2.1  The Graph Expression:

The HP 48 is capable of storing a large number (up to several hundred—limited only by available memory) of expressions for graphing and other calculations. In the introductory example, you selected the expression $\sin x$ by entering the expression and using the $\boxed{\text{STEQ}}$ key to store it for the graph. More commonly, you will enter such expressions and then *name* them with a unique name. Then you can select any expression for graphing by means of its name, without having to re-enter the expression or affecting any other stored expression.

When you execute DRAW to make a graph, the HP 48 looks at the contents of the special variable EQ to decide what to graph. If EQ contains a single expression (as in the example), that expression is graphed. However, if you store a new expression in EQ, the previous one is discarded. For this reason, the HP 48 also lets you store an expression in any other variable, then select the expression for graphing by storing the variable's *name* in EQ. Then when you change graphing expressions by storing a new name in EQ, you don't overwrite the stored expression itself.

The process of entering a new expression for graphing involves these steps:

(1)  Enter the expression.

(2)  Name it.

(3)  Select the expression's name for graphing.

"Enter the expression" in HP 48 terms means to place the expression in stack level 1. You can create the expression by pressing $\boxed{'}$ and typing the expression in a linear format between the ' ' characters in the command line. Or you can press $\boxed{\leftarrow}$ $\boxed{\text{EQUATION}}$ to enter the expression in a two-dimensional form. With either method, pressing $\boxed{\text{ENTER}}$ completes the entry and places the new object in stack level 1.

After entering a new expression, you can name it and select it for graphing (store its name in EQ) in a single operation. To illustrate, re-enter the $\sin x$ expression, this time with a name. First, type $'\text{SIN(X)}'$. Then press [←] [PLOT] [NEW], and you will see this display:

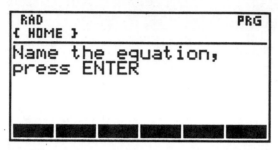

The HP 48 is prompting you to enter a name. Notice that the $\alpha$ annunciator is on, so you don't need to press [α] before typing the name characters. Type [S] [I] [N] [F] [ENTER]; now you should see

The second line of the display shows you that the "current equation" is $'\text{SIN(X)}$, which is stored with the name SINF. This time, variable EQ contains the name SINF rather than the actual graph expression. The term *current equation* is used by HP manuals to refer to the object that is selected for graphing (and for solving with HP Solve, for which the term *current equation* makes more sense), regardless of whether it is an expression, an actual equation, or even a program that is equivalent to an expression.

To add a second expression for graphing, just repeat the process outlined above. For example, enter the following:

$'\text{SIN(X)}^2$ [←]  [PLOT]  [NEW]  [SIN2]  [ENTER]

The new expression has replaced the previous choice SINF as the current equation. To reselect SINF, you can execute $'\text{SINF}'$ [STEQ]. The command STEQ is equivalent to $'\text{EQ}'$ STO; it stores an object in the variable EQ.

Once you have accumulated several graphing expressions, the most convenient way to select among them for graphing is to use the equation catalog. The catalog is activated by the $\boxed{\text{CAT}}$ key in the $\boxed{\text{←}}$ $\boxed{\text{PLOT}}$ menu:

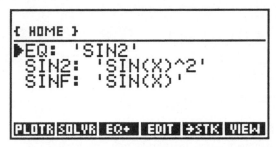

The catalog lists all of the variables that contain names or expressions, if any. In the example, you can see the SINF and SIN2 variables, as well as EQ itself. The contents of EQ shows which variable is currently selected for graphing. To graph a different expression, move the pointer to the desired variable using $\boxed{\triangle}$ or $\boxed{\triangledown}$, then press $\boxed{\text{PLOTR}}$. This stores the name you selected in EQ, then exits from the catalog and activates the menu containing the graphing commands ERASE and DRAW. You can also exit from the catalog without changing anything by pressing $\boxed{\text{ON}}$.

**Other Types of Graph Expressions.** In the examples thus far we have been considering graphing expressions that are simple functions of one variable. However, the HP 48 is capable of dealing with a wider class of "expressions:"

(1) **Multi-variable expressions.** HP 48 expressions can contain any number of variables, e.g. $'SIN(A+B*X)^C'$. To graph such an expression, you must designate one of the variables as the independent variable (see Section (9.2.3)), and supply values for the remaining variables so that the expression evaluates to a number. In this example, you might choose X as the independent variable; then you must store numbers in the variables A, B, and C before graphing. (You can actually store other expressions, or programs, in these variables, as long as they evaluate to numbers.) Storing 5 in A, 3 in B, and 2 in C, means that graphing the expression will draw a curve corresponding to $SIN(5+3*X)^2$.

(2) **Definition Equations.** It is common to speak of graphing $y = f(x)$, where you actually graph $f(x)$, while labeling the vertical axis $y$. If you select an equation of the form *name=expression*, where *name* is any variable name, then the HP 48 graphs as if it were just the expression $f(x)$. If you label the axes using LABEL, the vertical axis is labeled with *name*.

(3) **Other Equations.** When the current equation is actually a full *equation*, that is, two expressions combined with an "=" sign, the HP 48 graphs the two sides of the equation as separate curves. This approach derives from the connection between the graphing and solving systems; the values of the independent variable at which the two curves intersect are the solutions to the equation. This provides a simple method of graphing two expressions at once; you can form an equation by setting the two expressions equal to each other, then graph the equation.

**9.2.2  Plotting More than One Expression:** The HP 48 allows you to plot two or more expressions in the same operation, by combining the expressions into a *list* (yet another HP 48 data type). An easy way to do this is to enter all of the expressions onto the stack, then combine them with the → LIST operation. For example,

'SIN(X)'  [ENTER]  'COS(X)'  [ENTER]  '(X/3)^2 − 1'  [ENTER]

places three expressions on the stack. Then press [△][△][△] [→LIST] [ENTER] to combine the three into a list, indicated by the surrounding braces { }. Now name the list and select it for plotting:

[←] [PLOT]      [NEW]      LIST.EQ      [ENTER]

Notice that [NEW] entered the ".EQ" for you; this extension to the name ensures that the function catalog will include the list even though it is not an ordinary single expression.

If you now press [PLOTR] [ERASE] [DRAW], you will obtain this picture of all three expressions:

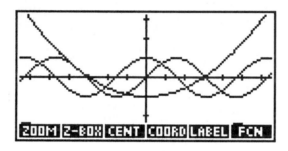

**9.2.3  Specifying the Graphing Variables:** In a graph of $y = f(x)$, $x$ is considered to be the *independent* variable. $y$ is the *dependent* variable, computed from successive values of $x$ by means of the function $f(x)$. The independent variable is represented by values along the abscissa, and the dependent variable by values along the ordinate.

In expressions that you enter for graphing, the HP 48 assumes by default that the independent variable is named X. (Note that the HP 48 is case sensitive: x is not the same as X.) However, there may be occasions where you wish to select a different variable as the independent variable. The command INDEP is provided in the [→] [PLOT] menu for this purpose. Entering ' *name* [INDEP] designates *name* as the new independent variable. It will remain so indefinitely, even when you change graphing expressions, until you select a new name.

For a function graph, the dependent variable does not appear in the graph expression except when the expression is a defining equation as explained in Section (9.21). However, for the purpose of labeling a graph, you can specify a dependent variable name, by using ' *name* ' DEPND. If you have not supplied a dependent variable name, the HP 48 uses Y.

**9.2.4  Plotting Scale and Viewing Window:** The HP 48 LCD contains 64 rows of 131 picture elements, or *pixels*. The correspondence between pixel positions and the logical coordinate system associated with a function graph is called the plotting scale. You establish the scale by assigning coordinates to the edges of the screen. These coordinates then also specify the ranges of the independent and dependent variables that are visible on a graph.

The HP 48 default for the ranges are −6.5 to +6.5 in the horizontal direction, and −3.1 to +3.2 in the vertical direction. These choices give a uniform scale in both directions of 0.1/pixel. The "tick marks" on the axes mark every tenth pixel; with the default ranges, the ticks are 1 unit apart in both directions.

You can change the scales manually by using the XRNG and YRNG commands, which are available as menu keys in the $\boxed{\text{r}}$ $\boxed{\text{PLOT}}$ menu. Each of these commands uses two numbers to set the coordinates of the pixels at the edges of the screen. For example, −10 10 $\boxed{\text{XRNG}}$ makes the $x$-coordinate of the leftmost pixel column to be −10, and the rightmost column to be +10. Similarly, 0 25 $\boxed{\text{YRNG}}$ sets the $y$-coordinate of the bottom row of pixels to be 0, and the top row to be 25. The RESET operation ($\boxed{\text{r}}$ $\boxed{\text{PLOT}}$ $\boxed{\text{NXT}}$ $\boxed{\text{RESET}}$) erases the current plot, and restores the default plotting scales and ranges.

In many cases you know the domain of an expression that you want to graph, but are uncertain about the appropriate vertical range. The AUTO command (next to DRAW in the $\boxed{\text{r}}$ $\boxed{\text{PLOT}}$ menu) can help: it samples the current equation at 40 points across the horizontal domain, then chooses a vertical range to include the maximum and minimum values computed, as well as the horizontal axis. AUTO then executes DRAW. This process can't always produce entirely satisfactory results, but it usually allows you to see enough of a curve on the screen to then correct the vertical range using the zoom features (Section 9.3.2) or the manual methods described in the preceding paragraphs.

The plotting ranges are often called the *viewing window*, with the idea that you are looking at an expression through a window that only lets you see a portion of the full domain and range of the expression. The HP 48 can also draw graphs that are larger than the actual LCD, so we need to distinguish between the viewing window imposed by the physical limits of the LCD, and a possibly larger *plotting window* that is the full range of rows and columns that you have allocated for plotting. For example, consider plotting the expression $'X*SIN(5*COS(X))'$ :

$'X*SIN(5*COS(X))'$     $\boxed{\text{↰}}$ $\boxed{\text{PLOT}}$ $\boxed{\text{NEW}}$   SINC $\boxed{\text{ENTER}}$    $\boxed{\text{PLOTR}}$ $\boxed{\text{NXT}}$ $\boxed{\text{RESET}}$    $\boxed{\text{↰}}$ $\boxed{\text{PREV}}$ $\boxed{\text{DRAW}}$

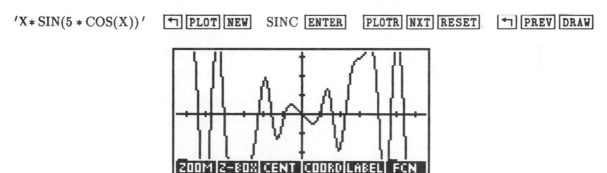

After a RESET, the plotting and viewing windows are the same size (131 × 64). With this choice, some of the maxima and minima are too high or low to be visible. You can create a larger graph without reducing the scale by using PDIM, which fixes the size of the plotting window. This command uses two arguments that represent the coordinates of the upper-left and lower-right corners of the new plotting window, preserving the current plotting scale (logical units/pixel). Let's stretch the plotting window to range from −8 to 8 in both directions:

$\qquad$ (−8, −8 )     (8, 8)     $\boxed{\text{r}}$ $\boxed{\text{PLOT}}$    $\boxed{\text{↰}}$ $\boxed{\text{PREV}}$    $\boxed{\text{PDIM}}$    $\boxed{\text{NXT}}$    $\boxed{\text{DRAW}}$.

The picture looks the same as before. But now press $\boxed{\text{↰}}$ $\boxed{\text{GRAPH}}$, then press and hold $\boxed{\triangle}$ until the picture stops moving:

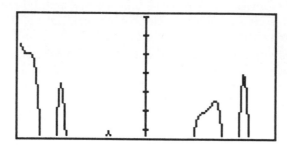

You can similarly scroll the viewing window in any of the four directions by pressing the cursor arrow keys. To restore the normal plot menu, press ⬅ GRAPH again.

**9.2.5 Positioning Axes:** By default, the HP 48 draws axes through (0,0). You can instruct the calculator to draw axes through any point you choose, by using the coordinates of such a point as the argument for AXES (➡ PLOT NXT NXT AXES). For example, (10,25) AXES stores (10,25) as the intersection of any axes that are drawn subsequently.

Due to the limited resolution of a calculator LCD, the HP 48 does not let you change the spacing of the tick marks on the axes. They are positioned at every tenth pixel, and serve primarily to distinguish the axes from other straight lines that might be present in a graph.

## 9.3 Working in Interactive Plot Mode

As you saw with the examples in the previous section, when you create a graph with DRAW or AUTO, the new graph remains visible, along with a menu of interactive plotting operations. This state is called the *interactive-plot mode*, where the display and keyboard are dedicated to plotting. You can exit from this mode at any time to return to ordinary calculator operations by pressing ON. This does not remove the current graph from memory; you can reactivate the interactive plot mode by pressing ◁ (when there is no command line).

We will illustrate the use of interactive plot mode with a function plot of $\sin(3\cos x)$:

(1)  **Set calculator modes.** Press ⬅ MODES STD to select the standard display format, then ⬅ RAD (if necessary) so that RAD appears in small characters at the upper left corner of the display.

(2)  **Enter and name the function.** ′SIN(3 ∗ COS(X))′ ⬅ PLOT   NEW   MODEX   ENTER.

(3)  **Select the *function* plot type (if necessary).** PTYPE FUNC.

(4)  **Reset the plotting ranges.** PLOTR   NXT   RESET.

(5)  **Graph the function.** ⬅ PREV   ERASE   DRAW.

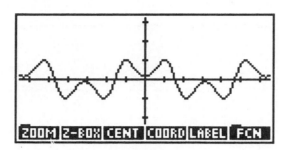

When the plotting is finished, the plot menu labels appear over the bottom row of the display. At any time, you can uncover the hidden part of the graph by pressing $\boxed{-}$ (to "take away" the labels), or $\boxed{\text{KEYS}}$ in the third page of the menu. To restore the labels, press any menu key, or $\boxed{+}$. You can also move the cursor downwards with $\boxed{\nabla}$ or $\boxed{\rightarrow}\boxed{\nabla}$. When the cursor reaches the top of the menu labels, further downward motion causes the graph to scroll upwards, revealing the hidden part.

Along with the menu keys (including $\boxed{\text{NXT}}$, $\boxed{\leftarrow}\boxed{\text{PREV}}$ and $\boxed{\rightarrow}\boxed{\text{PREV}}$), the rest of the HP 48 keyboard is redefined for plot-related operations:

(1)  $\boxed{\triangle}$, $\boxed{\triangleleft}$, $\boxed{\nabla}$, and $\boxed{\triangleright}$ move the plot cursor in the indicated directions. Pressing $\boxed{\rightarrow}$ before any of the arrow keys moves the cursor to the edge of the screen in that direction.

(2)  $\boxed{\text{ENTER}}$ returns the cursor position to the stack as a complex number.

(3)  $\boxed{\rightarrow}\boxed{\text{CLR}}$ erases the entire graph screen.

(4)  $\boxed{\times}$ sets the plot mark.

(5)  $\boxed{-}$ turns the menu labels on and off.

(6)  $\boxed{+}$ turns the coordinate display on and off.


**9.3.1  The Plot Cursor:** Hidden in the display picture above is the *plot cursor*, which is at the center of the screen, covered by the axes. To reveal the cursor, you can move it away from the axes with $\boxed{\triangleright}$ and $\boxed{\triangle}$:

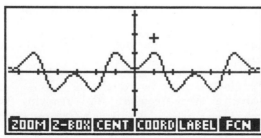

Now you can see the cursor, a small "+". The cursor is used as a pointer for various interactive operations. For example, if you press $\boxed{\text{COORD}}$ or $\boxed{+}$, you can read the cursor position in the menu label area:

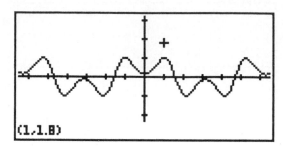

If you move the cursor while its coordinates are being displayed, the coordinates are changed along with the cursor position (this slows cursor movement down somewhat). You can restore the menu labels by pressing ⊞, ⊟, or any menu key.

You can copy the cursor position to the stack when you want to use its coordinates for further calculations, by pressing ENTER. You will see the busy annunciator flash momentarily, but there is no other visible indication until you exit interactive plot mode by pressing ON:

```
RAD
{ HOME }
4:
3:
2:
1:                    (1,1.8)
ERASE DRAW AUTO XRNG YRNG INDEP
```

Here you can see the cursor coordinates represented as a complex number. To return to interactive plot mode now, press ◁.

For some operations you must specify two points on the screen. The cursor is always available as one point; for the other, you set a *mark*. The mark is set by moving the cursor to a point, then pressing ✕ or MARK (in the third page of the plot mode menu). This places a ✕ mark on top of the cursor that is left behind when you move the cursor. If a mark already exists, the old mark is erased and a new one drawn—unless the cursor is actually on the mark, which just erases the mark.

If you press any menu key that requires a mark, when no mark exists, that key will set the mark rather than executing its labeled operation.

**9.3.2 Recentering and Zooming:** Very frequently while studying the graph of a function you may wish to change the plot ranges and scale to examine different features of the graph. The HP 48 provides keys in the plot menu for this purpose.

Recentering is a simple one-step operation by which you can move a graph relative to the screen so that some desired feature is centered on the screen. Just move the cursor to the point where you want the new screen center, then press CENT. For example, with the cursor as shown here

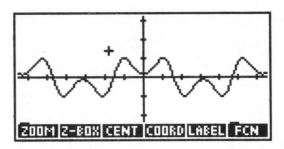

[CENT] redraws the picture like this:

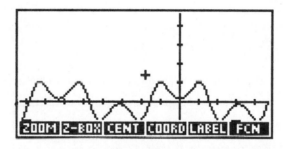

*Zooming* refers to the process of changing plot scales to magnify or demagnify a portion of a graph. On the HP 48, you may either enter numerical magnification factors, or, for zooming in, select a rectangular region that you want to expand to fill the screen. The latter is the easiest method. You position the mark and the cursor to define the region, then press [Z-BOX] (*Zoom-to-BOX*). This erases the screen and redraws the graph so that the original region fills the screen. For example, with the cursor and mark as shown,

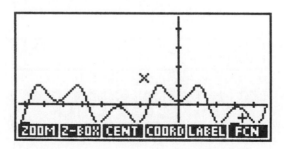

Z–BOX yields this picture:

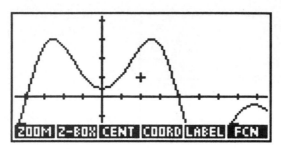

You can use $\boxed{\text{Z-BOX}}$ to change the scale in only one direction by placing the mark and the cursor in the same row or column. That is, if the mark and the cursor are in the same column, $\boxed{\text{Z-BOX}}$ increases the $y$-scale to stretch the marked range to fit the screen vertically, but leaves the $x$ – scale unchanged. Similarly, using $\boxed{\text{Z-BOX}}$ with the cursor and the mark in the same row increases the $x$ – scale but does not change the $y$ – scale.

When you want to zoom *out* so that more of a graph fits on the screen, you must use the numerical ZOOM operation. You may specify a magnification factor by which to multiply the $x$- or $y$ – scale, or both. A factor larger than 1 zooms *out*, shrinking plot features by that factor. A factor smaller than 1 zooms *in*, increasing the size of plot features.

With the picture as shown above, pressing $\boxed{\text{ZOOM}}$ changes the menu as shown here:

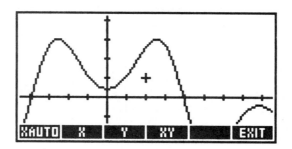

$\boxed{\text{EXIT}}$ allows you to cancel the change, returning to the main interactive plot mode menu without altering the plot. Pressing any of the other four labeled menu keys prompts you for a zoom magnification factor. For instance, pressing $\boxed{\text{X}}$ produces this display:

```
RAD                        PRG
{ HOME }

x axis zoom.
Enter value (zoom out
if >1), press ENTER



████  ████  ████  ████  ████  ████
```

Now you may type in a zoom factor (or press $\boxed{\text{ON}}$ to return to the zoom menu). Try 2 $\boxed{\text{ENTER}}$:

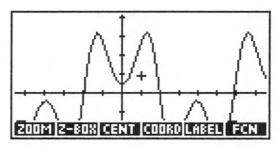

The plot is redrawn, with a doubled $x-$ scale. The logical coordinates of the center of the screen are preserved, so that the curve appears to have been compressed along the $x-$ direction, around the screen center. An analogous behavior is produced by pressing $\boxed{Y}$ in the zoom menu, with the $y-$ scale varying according to the zoom factor and the $x-$ scale held constant. $\boxed{XY}$ produces a uniform rescaling in both directions by the entered factor. The latter is often the most useful when you are searching for a particular plot feature; it lets you zoom out, repeatedly if necessary, until you can see the feature. Then you can zoom back in to magnify that area of the graph.

The final option in the zoom menu is $\boxed{\text{XAUTO}}$. This operation rescales in the $x-$ direction according to the zoom factor you enter, then autoscales in the $y-$ direction.

### 9.3.3   The Function Menu:
The $\boxed{\text{FCN}}$ menu key in interactive plot mode activates a sub-menu of operations that are applicable to function-type graphs. In general, the operations are designed for extracting information from a function curve at one or more points, where the points in question are specified by means of the cursor.

To help in exploring the operations, we will use the list of three functions used as an example in Section 9.2.2. If you did not enter that list, you should do so now. To select the list for plotting, press $\boxed{\Leftarrow}$ $\boxed{\text{PLOT}}$ $\boxed{\text{CAT}}$, use $\boxed{\triangle}$ or $\boxed{\triangledown}$ to move the selection pointer to LIST.EQ, and press $\boxed{\text{PLOTR}}$. Then press $\boxed{\text{NXT}}$ $\boxed{\text{RESET}}$ $\boxed{\Leftarrow}$ $\boxed{\text{PREV}}$ $\boxed{\text{ERASE}}$ $\boxed{\text{DRAW}}$ to make the plots:

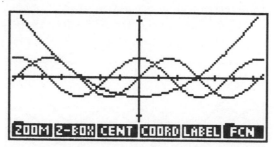

When you are working with a single plot function, all of the function menu operations are directed to that function. If you are working with a list of functions, as in this example, the menu operations apply to the first function in the list; ISECT may also use the second object. To determine which object is the first in the list, you can either

(1)   press $\boxed{\Leftarrow}$ $\boxed{\text{REVIEW}}$, which shows the beginning of the equation list, which is usually enough to identify the first object; or

(2)   (in the FCN menu) press $\boxed{\text{NXT}}$ $\boxed{\text{F(X)}}$. This displays the value of the selected object at the horizontal position of the cursor, and moves the cursor to the selected curve.

In the current example, moving the cursor a few pixels to the right, then pressing $\boxed{\text{F(X)}}$, moves the cursor to the $'\text{SIN(X)}'$ curve:

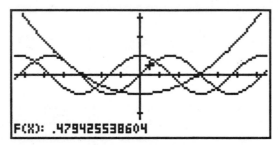

On the same page with $\boxed{\text{F(x)}}$ is $\boxed{\text{NXEQ}}$, which selects the next object in the current equation list:

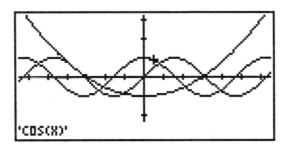

Notice that $\boxed{\text{NXEQ}}$ also displays the newly selected object at the bottom of the screen. Using this operation, you can cycle through all of the objects in the current equation list. It works by actually moving the first object to the end of the list, so that the second object moves to the front.

Returning ($\boxed{\text{NXT}}$) to the first page of the menu, the first entry there is $\boxed{\text{ROOT}}$, which finds a zero of the selected function. In the current example, pressing $\boxed{\text{ROOT}}$ finds the zero of $\cos(x)$ at $\pi/2$:

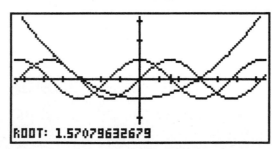

ROOT invokes a numerical root-finder on the selected function, using the horizontal cursor position as an initial guess. When it has found a root, it displays the value at the bottom of the screen and moves the cursor to the root. You can press any key to restore the menu.

If you next press $\boxed{\text{SECT}}$, you will see the cursor move to an intersection of two curves:

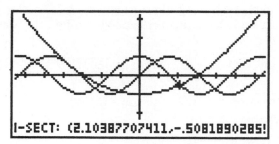

ISECT finds the intersection of the first two curves obtained from the current equation list:

(1) When the entry is actually an equation, ISECT finds the intersection of the two expressions that comprise the left and right sides of the equation; ISECT is the same as ROOT in this case.

(2) When neither of the first two entries are equations, ISECT finds their intersection.

(3) When the first entry is an expression, and the second is an equation, ISECT substitutes the first expression for the left side of the equation, and solves the resulting equation.

The next key in the function menu is $\boxed{\text{SLOPE}}$. This operation computes the slope of the selected curve at the $x -$ position of the cursor:

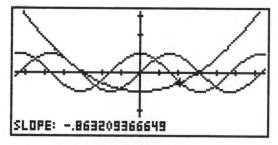

The slope is computed from the symbolic derivative of the selected function. Differentiation is also used by EXTR, which finds the critical points of a curve. Pressing $\boxed{\text{EXTR}}$ here finds the local minimum of the cosine curve at $\pi$:

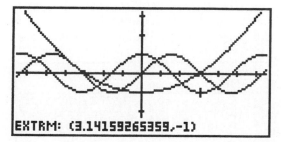

EXTR differentiates the selected function, then uses the root-finder to find a critical point where the derivative is zero.

AREA uses the HP 48's numerical integration capability to find the area of a curve between two $x$ – values selected by the cursor and the mark. For example, to find the area of the "hump" of the cosine curve between its zeros at $\pm\pi/2$ :

(1)    Move the cursor near to the zero at $\pi/2$, and press $\boxed{\text{ROOT}}\ \boxed{\times}\ \boxed{\times}$ (the first $\boxed{\times}$ just restores the labels).

(2)    Move the cursor near to the zero at $-\pi/2$, and press $\boxed{\text{ROOT}}$.

(3)    Press any key, then $\boxed{\text{AREA}}$ :

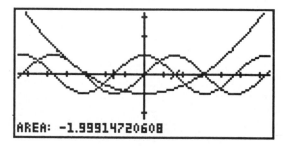

The difference between this result and the ideal result –2 (negative because you integrated from right to left) arises primarily because the integration limits are the $x$ – coordinates of the centers of the pixels near the zeros, rather than the computed coordinates.

The remaining FCN menu operation is F $'$ (on the second page). This operation adds the derivative of the selected object to the head of the current equation list, then erases the screen and executes DRAW again. For example, press $\boxed{\text{NXEQ}}$ to select the parabola, then $\boxed{\text{F}'}$ :

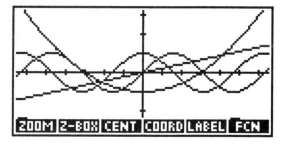

The derivative of the parabola is a straight line with slope 0.5.

If you exit to the stack at this point ($\boxed{\text{ON}}$), you will see a display like this:

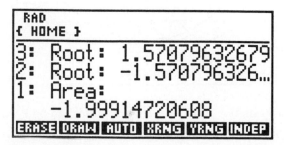

All of the function menu operations that compute results for display in interactive plot mode also return those results to the stack. Each result is tagged with the operation name to help identify it.

## 9.4 Polar Graphing

The HP 48 is capable of making graphs in polar coordinates directly from an equation for the radius as a function of the polar angle. The steps in this process are as follows:

(1) Select the POLAR plot type: press $\boxed{\text{⬑}}$ $\boxed{\text{PLOT}}$ $\boxed{\text{PTYPE}}$ $\boxed{\text{POLAR}}$.

(2) Enter the function $r = f(\theta)$, or just $f(\theta)$, using the same method as outlined in Section 9.2.1. You can use any name you want for the polar angle; to type a $\theta$, press $\boxed{\alpha}$ $\boxed{\text{⮕}}$ $\boxed{\text{F}}$.

(3) Set the screen ranges as described in Section 9.2.4.

(4) Select the independent variable, in this case the polar angle. Press $\boxed{\text{PLOTR}}$ or $\boxed{\text{⮕}}$ $\boxed{\text{PLOT}}$, then type the quoted variable name and press $\boxed{\text{INDEP}}$.

(5) Set the range of the polar angle by entering the starting value and the ending value, then pressing $\boxed{\text{INDEP}}$ again.

(6) Press $\boxed{\text{ERASE}}$ (if desired), then $\boxed{\text{DRAW}}$.

**Example 1.** Spiral $r = \theta/2\pi$, for $0 \le \theta \le 6\pi$

(1) Press $\boxed{\text{⬑}}$ $\boxed{\text{PLOT}}$ $\boxed{\text{PTYPE}}$ $\boxed{\text{POLAR}}$ (if necessary).

(2) Enter $'R = \theta/(2 * \pi)'$ $\boxed{\text{NEW}}$ SPIRAL $\boxed{\text{ENTER}}$.

(3) Press $\boxed{\text{⮕}}$ $\boxed{\text{PLOT}}$ $\boxed{\text{NXT}}$ $\boxed{\text{RESET}}$ $\boxed{\text{⬑}}$ $\boxed{\text{NXT}}$.

(4) Enter $'\theta'$ $\boxed{\text{INDEP}}$.

(5) Enter 0 $\boxed{\text{⬑}}$ $\boxed{\pi}$ $\boxed{\text{⮕}}$ $\boxed{\rightarrow\text{NUM}}$ 6 $\boxed{\times}$ $\boxed{\text{INDEP}}$.

(6) Press $\boxed{\text{ERASE}}$ $\boxed{\text{DRAW}}$ $\boxed{-}$:

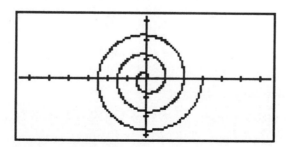

**Example 2.** Limaçon $r = 1.5(1 - 2\cos\theta)$.

(1) Press $\boxed{\text{⬑}}$ $\boxed{\text{PLOT}}$ $\boxed{\text{PTYPE}}$ $\boxed{\text{POLAR}}$ (if necessary).

(2) Enter $'R = 1.5 * (1 - 2 * \text{COS}(\theta))'$    NEW    LIMAÇON ENTER. (You can enter a Ç by pressing C ← 9).

(3) Press ← PLOT, then, if necessary to reset the default plotting ranges, NXT    RESET    ←  NXT.

(4) Enter $'\theta'$  INDEP.

(5) Enter 0    ← π 2 ×    ← → NUM    INDEP.

(6) Press ERASE    DRAW    — :

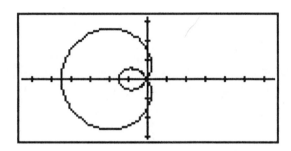

## 9.5  Parametric Graphing

The polar graphing described in the previous section can be considered as a special case of *parametric graphing*, where the $x$- and $y$-coordinates of plotted points are expressed as separate functions of a third parameter $t$. For polar graphs, $T$ is the polar angle $\theta$, and the functions are

$$x = r \cdot \cos\theta, \qquad y = r \cdot \sin\theta.$$

For parametric graphing on the HP 48, the two functions are combined into a single expression in $t$ that returns a complex number when evaluated. A complex number is represented as an ordered pair $x, y$. The pair of parametric equations $x = f_x(t)$ and $y = f_y(t)$ can be combined into a complex-valued expression suitable for parametric graphing as $f_x(t) + i \cdot f_y(t)$.

The steps in the parametric graphing process then are as follows:

(1) Select the PARAMETRIC plot type: press ← PLOT PTYPE PARA.

(2) Enter the function $f_x(t) + i \cdot f_y(t)$ using the same method as outlined in Section 9.2.1. You can use any name you want for the independent variable $t$.

(3) Set the screen ranges as described in Section 9.24.

(4) Identify the independent variable. Press PLOTR or ← PLOT, then type the variable name and press INDEP.

(5) Set the range of the independent variable by entering its starting value and its ending value, then press INDEP again.

(6) Press ERASE (if desired), then DRAW.

**Example 1:** Lissajous figure $x = 3\sin 3t$, $y = 2\sin 4t$, for $\theta \leq t \leq 6.5$.

(1)  Press [←] [PLOT] [PTYPE] [PARA] (if necessary).

(2)  Enter $'3 * SIN(3 * T) + 2 * i * SIN(4 * T)'$   [NEW]   LISSAJOUS   [ENTER].

(3)  Press [→] [PLOT]   [NXT]   [RESET]   [←] [NXT].

(4)  Enter $'T'$   [INDEP].

(5)  Enter 0   6.5   [INDEP]

(6)  Press [ERASE]   [DRAW]:

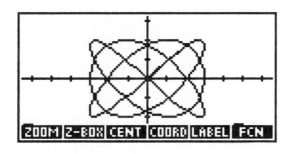

## Bibliography

The following books provide additional information on the operation and methodology of the HP 48S and HP 48SX calculators.

Coffin, Coffin, and Loux. *An Easy Course in Using the HP 48SX.* Grapevine Publications (Corvallis OR, 1990). An introduction to basic HP 48 operation.

Depew, R. Ray. *HP 48 Graphics.* Grapevine Publications (Corvallis, 1991). How to create custom display images with diagrams, pictures, icons, labels, titles, multiple plots, games and menus.

Hewlett-Packard Co. *HP 48 Programmer's Reference Manual.* Hewlett-Packard Co. (Corvallis 1990). Concise reference information for the HP 48, including a brief alphabetical listing of programmable keywords, units, flags, reserved variables, and error messages.

Wickes, William C. *HP 48 Insights, Part I.* Larken Publications (Corvallis, 1991) In-depth treatment of the principles and programming of the HP 48.

Wickes, William C. *HP 48 Insights, Part II.* Larken Publications (Corvallis, 1991) Exposition of the HP 48's pre-programmed resources, including graphing, solving, unit managment, symbolic algebra and calculus, and statistics.

# Chapter 10

## Apple II Version of *Master Grapher*

### 10.1  Starting Up

Remove the *Master Grapher* disk from its envelope and insert it in disk drive #1 with the label up and the oval slot forward (see Figure 10.1).

**Figure 10.1.** Insert *Master Grapher* disk in disk drive #1.

Carefully push the disk all the way into the disk drive and then close the gate on the drive until it clicks into place. Turn on the computer and the monitor. *Master Grapher* will automatically run and the title page will be visible. Press any key to continue to the **Main Menu** (see Figure 10.2) to select the graphing program you wish to use.

```
Please choose one:

1) Run the Function Grapher
2) Run the Conic Grapher
3) Run the Parametric Grapher
4) Run the Polar Grapher
5) Run the 3D Surface Grapher

6) Run the Instructional Grapher
7) See instructions

Press the key of your selection (1-7): ▮
```

**Figure 10.2.** Main Menu showing available graphing programs and instructions.

Select the graphing program of your choice by pressing the numbers 1 through 5. Choice # 6 is a self-running demonstration of the **Function Grapher** program. Choice # 7 provides written instructions about the operation of the programs.

## 10.2  Apple II Function Grapher

Select #1 from the **Main Menu** to run the **Function Grapher** program. After a short pause while the program loads, a title screen will appear (see Figure 10.3). This screen asks the question "Do you have a color monitor (Y/N)"? If you are using a color monitor, respond by keying $\boxed{\text{Y}}$ and if you are using any other type of monitor respond by keying $\boxed{\text{N}}$. Operation of the program will not be affected if you respond with a $\boxed{\text{Y}}$ and you are not using a color monitor. In some cases, specifying a color monitor when really using a green-screen monitor can help students see slight differences between functions. If you are making printouts of the screen, selecting $\boxed{\text{N}}$ may give better results.

> Function Grapher version 1.0
>
> Concept and design by Bert Waits and
>     Frank Demana
>
> Machine Language programming by Greg
>     Ferrar, Senior at Worthington High
>     School
>
> Copyright © 1986, 1987, 1988, 1989 by
>     Bert Waits and Frank Demana
>
> Do you have a color monitor (Y/N)? ▪

**Figure 10.3.** Title screen for the function grapher.

Immediately after entering the information about the type of monitor being used, the **graphing screen** will appear and the graph of the default function $f(x) = \sin(x)$ will be drawn. Figure 10.4 shows the **graphing screen** with the default **viewing rectangle** of $[-10, 10]$ by $[-10, 10]$. The values corresponding to the $x$-**minimum**, $x$-**maximum**, $y$-**minimum**, and $y$-**maximum** of the viewing rectangle are displayed on the screen in the corresponding places around the **graphing window**.

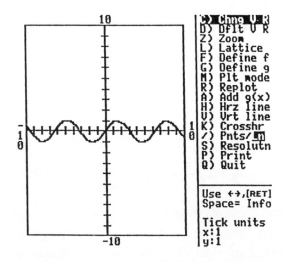

**Figure 10.4.** Graphing Screen showing the graph and the commands menu.

The right side of this screen contains the **commands menu**. Commands are selected by keying the letter or symbol listed for each command, or by using the arrow keys to slide the highlighter bar to a command and pressing $\boxed{\text{Return}}$. Below the **commands menu** is a window containing information about selecting commands and displaying information about the functions which are graphed. The information about **"Tick units"** in the lower right corner of the graphing screen indicates the distance between each tick mark on the horizontal ($x$) and vertical ($y$) axes. Pressing the $\boxed{\text{Space Bar}}$ while a graph is plotting will pause the graph and give the coordinates of the last point plotted (see Figure 10.5). Press any key to continue. Pressing the $\boxed{\text{Space Bar}}$ at any other time will display a window showing the last function or functions which were graphed. Press any key to return to the **commands menu** and erase the **function display window**. Figure 10.6 shows the screen with this function display window activated.

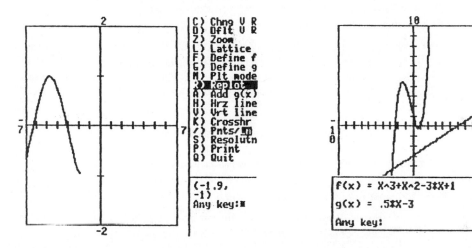

**Figure 10.5.** Graph paused.          **Figure 10.6.** Function information window.

The definition for the commands which appear in the menu at the right of the graphing screen are as follows:

**C ) Chng V R.** (Change Viewing Rectangle) This command is used to change the the visible viewing rectangle by entering new values for **x-min**, **x-max**, **y-min**, and **y-max**.

**D ) Dflt V R.** (Default Viewing Rectangle) This command will replot the graph(s) in the default viewing rectangle of $[-10, 10]$ by $[-10, 10]$. Graphs drawn depend on the **Plot mode** chosen (see **M) Plt mode** below).

**Z) Zoom.** This command activates the **Zoom Menu** which controls the zoom-in and zoom-out functions of the program.

**L) Lattice.** This command will overlay a lattice (an array of dots) in the viewing rectangle. The dots correspond to the tick marks on the two axes. The distance between the dots in the horizontal or vertical direction corresponds to the values for the "Tick units" found in the lower right corner of the screen.

**F) Define f.** Use this command to input a new function, $f$. This function will be plotted immediately in the default viewing rectangle unless **A ) Add f(x)** from the **Plot Options** menu is active.

**G)**    **Define g.** Use this command to input a second function, $g$. To display the graph of this second function, select **A ) Add g(x)**.

**P)**    **Plt mode** (Plot mode). This command allows the user to specify which graphs (ie. $f$, $g$, $f^{-1}$, $g^{-1}$) are automatically drawn on the screen. Once a mode is set, it remains active until changed by the user.

**R)**    **Replot.** This command replots the current function(s) in the current viewing rectangle. The current **Plot mode** (see **M) Plt mode** above) determines which graphs will be replotted.

**A)**    **Add g(x).** Select this command to overlay the function $g$ on the current screen. The function $g$ will be displayed along with any other functions currently displayed. You may add as many $g$ functions as you wish by using the **G ) Define g** and **A ) Add g(x)** commands to overlay graphs. Only the last g function remains in the memory of the computer.

**H )**    **Hrz line** (Horizontal line). This command places a moving horizontal line on the graphing screen. Use the $\boxed{\downarrow}$ $\boxed{\uparrow}$ keys, or the $\boxed{U}$ (up) and $\boxed{D}$ (down) keys to move this line to any position on the screen. Pressing the $\boxed{\text{Space Bar}}$ causes the $y$-intercept of the horizontal line to be displayed. Pressing the $\boxed{G}$ key allows the user to enter a $y$-coordinate for the horizontal line to "GOTO"; press $\boxed{\text{Return}}$ to put the horizontal line at the selected position. Press $\boxed{\text{Esc}}$ to return to the **commands menu**.

**V)**    **Vrt line** (Vertical line). This command places a moving vertical line on the graphing screen. Use the $\boxed{\leftarrow}$ $\boxed{\rightarrow}$ keys, or the $\boxed{L}$ (left) and $\boxed{R}$ (right) keys to move this line to any position on the screen. Pressing the $\boxed{\text{Space Bar}}$ causes the $x$-intercept of the vertical line to be displayed. Pressing the $\boxed{G}$ key allows the user to enter an $x$-coordinate for the vertical line to "GOTO"; press $\boxed{\text{Return}}$ to put the vertical line at the selected position. Press $\boxed{\text{Esc}}$ to return to the **commands menu**.

**K)**    **Crosshr** (Crosshairs). This command combines the moving vertical and horizontal lines so that the coordinates of any point on the screen can be estimated. Use the $\boxed{\leftarrow}$ $\boxed{\rightarrow}$ $\boxed{\downarrow}$ $\boxed{\uparrow}$ keys or the $\boxed{L}$ $\boxed{R}$ $\boxed{U}$ $\boxed{D}$ keys to move the two lines to the desired position. Press the space bar to display the coordinates of the intersection of the two lines (shown in the lower right hand corner of the screen). Press $\boxed{G}$ to specify a point for the crosshairs to "GOTO" and press $\boxed{\text{Return}}$ to send them to that point. Press $\boxed{\text{Esc}}$ to return to the **commands menu**.

**/ )**    **Pnts/Ln** (Points / Lines). This command selects one of two graphing modes, points or line segments. When the Pnts mode is selected, the program will only plot the points $(x, y)$ evaluated for a given function. When the Ln mode is selected, consecutive pairs of points will be connected with a line segment. The Ln mode is the default mode.

**S )**    **Resolution.** The resolution of the graph refers to the density of the displayed points. This command allows the user to increase or decrease the density of points in the domain of the function chosen to be evaluated. The result of increasing the density of points is to produce a more accurate graph. However, the time required to draw the denser graph will increase because of the increased number of points which are evaluated.

**P )**    **Print.** This command allows the user to print the current graph on paper. To produce a screen-dump of the entire screen including the graph and **commands menu**, press $\boxed{\text{Control}}$ $\boxed{P}$ (at the same time).

**Q )**    **Quit.** This command allows the user to exit the program by rebooting the disk. Use this option to return to the **Main Menu** to select another graphing program or to end the session.

## 10.3  Detailed Guide to Using the Interactive Menu Driven Commands

Use the instructions on **"Starting Up"** in Section 10.1 to run *Function Grapher*. The first graphing screen for *Function Grapher* will show the graph of the default function $f(x) = \sin x$ in the default viewing rectangle (see Figure 10.4).

**C)  Chng V R** (Change Viewing Rectangle). This command is used to change the portion of the coordinate plane visible in the graphing window. This visible portion of the plane is known as the **viewing rectangle** and is designated by the parameters **x-minimum**, **x-maximum**, **y-minimum**, and **y-maximum**. These parameters correspond to the horizontal and vertical dimensions of the viewing rectangle and are displayed in their appropriate positions along the edges of the viewing rectangle.

Suppose we want to view the graph of $f(x) = \sin x$ in the $[-7, 7]$ by $[-2, 2]$ viewing rectangle. To change the dimensions of the viewing rectangle, select the **C) Chng V R** command and the menu at the right of the viewing rectangle will be replaced with a prompt asking for a new value for **x-min**; enter a value for **x-min** and press $\boxed{\texttt{Return}}$. The next prompt will ask for a value for **x-max**; enter a value and press $\boxed{\texttt{Return}}$. Repeat the same procedure for values of **y-min** and **y-max**. After the last value is entered, the graph will be redrawn in the new viewing rectangle. Figure 10.7 shows the screen with the new viewing rectangle of $[-7, 7]$ by $[-2, 2]$ entered. Figure 10.8 shows the graph of $f(x) = \sin x$ redrawn in the new viewing rectangle.

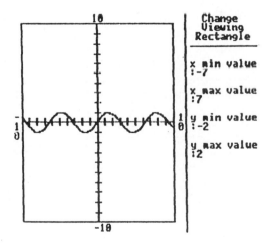

**Figure 10.7.** Enter new parameters.

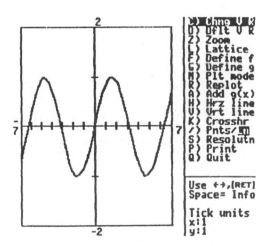

**Figure 10.8.** Graph redrawn.

If you make a mistake when entering a value, before pressing $\boxed{\texttt{Return}}$ press the $\boxed{\leftarrow}$ key to backspace and erase characters. Press the $\boxed{\texttt{Esc}}$ key to return to the **commands menu** without changing the current viewing rectangle.

**Z)  Zoom.** Select this option to activate the **Zoom Menu** to zoom in or zoom out on the graph. The **Zoom Menu** appears in place of the **commands menu** at the right of the graph and shows the four zoom options available in the program (see Figure 10.9). To select an option, key the corresponding letter or position the highlight bar and press $\boxed{\texttt{Return}}$.

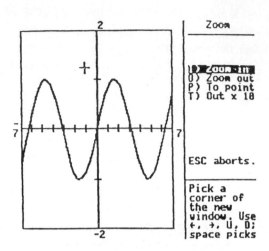

**Figure 10.9.** Select a corner.

**I)**  **Zoom in.**  This command allows you to make a *close-up* or *magnified* view of a portion of the graph by capturing that area in a rectangular box or window. Select **I ) Zoom in**; a small crosshair ( + ) will appear in the center of the screen and you will be asked to "Pick a corner of the new window." Move the crosshair to a corner of the new window using the $\boxed{\leftarrow}$ $\boxed{\rightarrow}$ $\boxed{\downarrow}$ $\boxed{\uparrow}$ keys or the $\boxed{L}$ $\boxed{R}$ $\boxed{U}$ $\boxed{D}$ (left, right, up, down) keys (see Figure 10.10) When the crosshair is in position, press the $\boxed{\text{Space Bar}}$ to set this corner. Using the arrow keys, stretch the new viewing window horizontally and vertically to capture the area of the graph you want to examine more closely (see Figure 10.10). When you are satisfied with the location of the viewing window, press the $\boxed{\text{Space Bar}}$ and the function will be replotted in the new viewing rectangle (see Figure 10.11). The

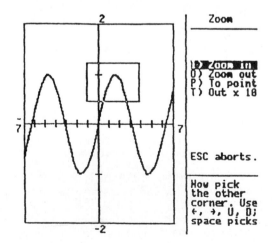

**Figure 10.10.** Select the diagonal corner.

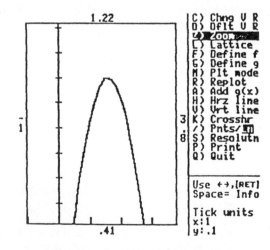

**Figure 10.11.** Graph redrawn in the new viewing rectangle.

values for **x-min**, **x-max**, **y-min**, and **y-max** around the edges of the screen now represent the new viewing rectangle established by the rectangular window. Press $\boxed{\texttt{Esc}}$ to cancel the zoom process and return to the **commands menu**.

**O)**  **Zoom out.** This command allows you to zoom-out, or back away from a graph to see a larger portion of the coordinate plane. The zoom-out is done relative to the center of the screen by the horizontal and vertical factors you specify. Select **O ) Zoom out** and the **Zoom Menu** will be replaced with a prompt asking "Please enter the zoom factor for the $x$-axis." Entering the value "5" will multiply the horizontal distance across the screen by 5. Enter a value and press $\boxed{\texttt{Return}}$. Next, the same prompt for the $y$-axis will appear; enter a value and press $\boxed{\texttt{Return}}$. After the two factors are entered, the graph will automatically be redrawn in the new viewing rectangle.

When you are not sure that the current graph is a complete graph of a function, you can use the **O ) Zoom out** command to quickly look for an appropriate viewing rectangle that displays a complete graph, or to determine that the original view was a complete graph. Often you will need to use different zoom factors for the $x$-axis and the $y$-axis as we did above.

*Note*  The **O ) Zoom out** command can be used to zoom in by setting the zoom factors between 0 and 1.

**P)**  **To point.** This command allows you to zoom in on a selected point of a graph by a factor of 0.1. Select **P ) To point** and a crosshair (+) will appear at the center of the screen. Use the $\boxed{\leftarrow}$ $\boxed{\rightarrow}$ $\boxed{\downarrow}$ $\boxed{\uparrow}$ keys to move the crosshair to the point you wish to be the center of the new viewing rectangle and press $\boxed{\texttt{Space Bar}}$. The graph will be redrawn using the selected point as the center of the new viewing rectangle and the distance across the new viewing rectangle will be 1/10 of the previous viewing rectangle (zoom-in by a factor of 0.1). Figure 10.12 shows the crosshair in position on the graph of $f(x) = \sin x$ in the viewing rectangle $[-7, 7]$ by $[-2, 2]$. This viewing rectangle represents a distance of 14 units in the horizontal and 4 units in the vertical direction. Figure 10.13 shows the graph redrawn in the new viewing rectangle $[.91, 2.31]$ by $[.81, 1.21]$ centered on the selected point. This viewing rectangle represents a distance of 1.4 units in the horizontal and .4 units in the vertical direction.

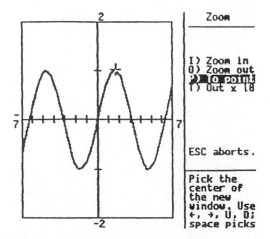

**Figure 10.12.** Center point selected.

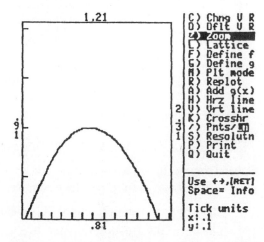

**Figure 10.13.** Zoom in by factor of 1/10.

**T) Out x 10.** This command zooms out by a factor of 10 relative to the center of the screen. The distance across the screen in the horizontal and vertical direction is multiplied by 10. The center of the new viewing rectangle is the same point as the center of the previous viewing rectangle.

**L) Lattice.** This command overlays a lattice (an array of dots) in the viewing rectangle. The distance between these dots is given by the **Tick units**, or scale values, found in the lower right-hand corner of the screen. This command is helpful when you want to estimate the coordinates of a point that is not on an axis. For example, select **D) Dflt V R** to draw the graph of $f(x) = \sin x$ in the default viewing rectangle and then zoom in on some portion of the graph using the **Zoom-in** commands. Next, select **L) Lattice** to obtain a lattice in the viewing rectangle. The **Tick units** specify the distance between any two dots vertically and horizontally. Now, it is easy to estimate the coordinates of some point on the graph.

**F) Define f.** This command allows you to enter a new function to be graphed. To enter the new function $f(x) = x^3 + x^2 - 3x + 1$, proceed as follows: select **F) Define f** and a window will appear showing the current $f$ function and a prompt asking you to enter the new $f$ function (see Figure 10.14).

We need to enter 3 and 2 as exponents. The Apple computer uses the usual BASIC language notations caret symbol $\wedge$ ( $\boxed{\text{Shift}}$ $\boxed{6}$ ) to indicate exponentiation and the symbol $*$ ( $\boxed{\text{Shift}}$ $\boxed{8}$ ) to indicate multiplication. (If you are using an Apple II +, use $\boxed{\text{Shift}}$ $\boxed{\text{N}}$ for $\wedge$ and $\boxed{\text{Shift}}$ $\boxed{:}$ for $*$.) Type in the keystrokes $\boxed{\text{X}}$ $\boxed{\wedge}$ $\boxed{3}$ $\boxed{+}$ $\boxed{\text{X}}$ $\boxed{\wedge}$ $\boxed{2}$ $\boxed{-}$ $\boxed{3}$ $\boxed{*}$ $\boxed{\text{X}}$ $\boxed{+}$ $\boxed{1}$ and press $\boxed{\text{Return}}$. The graph of $f(x) = x^3 + x^2 - 3x + 1$ will be drawn in the default viewing rectangle (unless the **A) Add f(x)** option is selected from the **Plot Options** menu). See Figure 10.15.

**Figure 10.14.** Enter the new $f$ function.

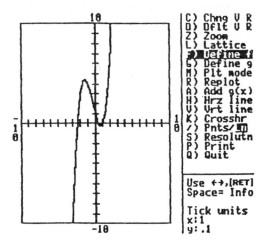

**Figure 10.15.** New function graphed.

Special keying instructions are needed to enter built-in functions or to obtain correct graphs of some special functions. Here are some examples.

(1) The absolute value function $|\mathbf{X}|$ is entered as **ABS ( X )**.

(2) The natural logarithm function $\ln x$ is entered as **LOG ( X )**. Note: $y = \log_b(x)$ can be graphed for any base b by entering **LOG ( X ) / LOG ( B )**.

(3)  The exponential function $e^x$ is entered as **EXP ( X )**.

(4)  The root function $\sqrt[n]{x}$ for $n$ **odd** is entered as **ABS ( X ) / X * ABS ( X ) ∧ ( 1 / N )**. You can enter $x^{1/n}$ as $x\char94(1/N)$, but for $n$ odd, you will only get the portion of the graph in the first quadrant.

(5)  The greatest integer function $[\mathbf{x}]$ is entered as **INT ( X )**.

(6)  The signum function is entered as **SGN ( X )**.

(7)  The square root function is entered as **SQR ( X )**.

**G)  Define g**.  This command, together with the **A ) Add g(x)** command, is used to overlay the graph(s) of additional function(s) in the same viewing rectangle with the current graph of the function $f$. Operation of this command is identical to the **F) Define f** command except that once a $g$ function is entered into the computer, its graph will not be drawn automatically; the **A ) Add g(x)** command must be used to draw the graph.

If you use the **G ) Define g** and **A ) Add g(x)** commands several times, then the graph of $f$ together with the graphs of all the $g$ functions will appear in the same viewing rectangle. Thus, you can overlay the graphs of as many functions as you wish. The program, however, only holds two function definitions at a time, the last $f$ and $g$ function defined; once a new $f$ or $g$ function is defined, it replaces any other function entered. If you replot the functions or change the scale of the viewing rectangle, the only functions available are the latest $f$ and $g$ functions entered. Figure 10.16 shows the screen for entering the $g$ function and Figure 10.17 shows the graph of $y = .5x - 3$ overlaid in the viewing rectangle using the **A ) Add g(x)** command. Note: The option **A ) Add f(x)** from the **Plot Options** menu can do much the same thing.

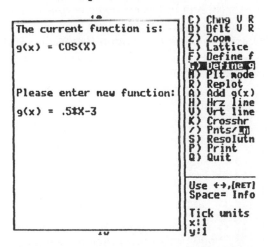

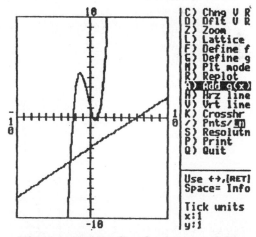

**Figure 10.16.** Enter new $g$ function.          **Figure 10.17.** Graph of $g$ overlaid.

**M)  Plot mode**.  Selecting this option activates the **Plot Mode Options Menu** (see Figure 10.18). This menu allows you to specify which functions you would like to plot automatically when graphs are redrawn. For example, selecting option **5 ) f and g**, causes both currently defined functions, $f$ and $g$, to be drawn when zooming in or out, or when the **R ) Replot** command is selected. By selecting **G ) g(x)** you could make the $g$ function the default function in all plotting formats. This menu also allows plotting of the inverse relationships $f^{-1}(x)$ and $g^{-1}(x)$ in various combinations. **F ) f(x)** is the default setting for the **Plot Mode**. If you select the **G ) g(x)** option, then you can overlay the graph of $f(x)$ by selecting the **A ) Add f(x)** option from the **Plot Options** menu and then **R ) Replot** from the **commands menu**. Once a new **Plot Mode** is selected it stays active until you change it again.

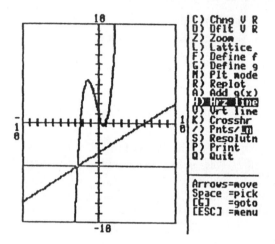

**Figure 10.18.** Plot Mode Options Menu.

**H) Hrz line.** This command is used to draw a moving horizontal line in the current viewing rectangle. The lower right hand corner of the screen contains directions on how to manipulate the line. Use the $\boxed{\downarrow}$ $\boxed{\uparrow}$ keys (or $\boxed{U}$ (up) and $\boxed{D}$ (down) keys) to move the line to the desired location (see Figure 10.19). (Note: the horizontal line first appears in the center of the screen; if the $x$-axis is in the center of the screen, the horizontal line will cover the axis and it will appear that the axis has disappeared. The axis

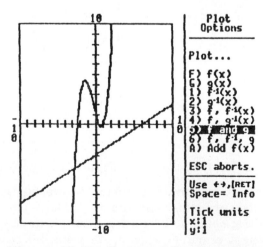

**Figure 10.19.** Horizontal line in position.

will reappear when the line is moved.) Press the $\boxed{\texttt{Space Bar}}$ to get the $y$-intercept of the line. Press $\boxed{\texttt{G}}$ for "GOTO" to send the line to a specific location in the viewing rectangle. Press $\boxed{\texttt{Esc}}$ to return to the **commands menu**.

**K) Crosshr.** This command draws both a moving horizontal and a moving vertical line in the viewing rectangle. Operation of this command is like the **H ) Hrz line** and **V ) Vrt line** commands. This command can be used to estimate the coordinates of any point in the viewing rectangle by placing the intersection of the horizontal and vertical lines on that point and pressing the $\boxed{\texttt{Space Bar}}$. Figure 10.20 shows the crosshair positioned at a local maximum value of the graph of $f(x) = x^3 + x^2 - 3x + 1$. The estimate of that point is $(-1.5, 4.4)$ as seen in the lower right-hand corner of the screen. Press $\boxed{\texttt{G}}$ to GOTO a specific point in the viewing rectangle. The program will ask for $x$ and $y$ values to locate the crosshair. Press $\boxed{\texttt{Esc}}$ to return to the **commands menu**.

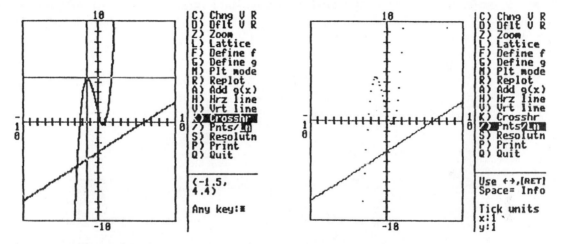

**Figure 10.20.** Crosshair on a local maximum.          **Figure 10.21.** Graph drawn as points.

**/ ) Pnts/Ln.** This command gives you the option of drawing a graph by only plotting individual points or by connecting the points with straight line segments. The default setting is **/ ) Pnts/Ln,** meaning that graphs will be drawn using line segments to connect each pair of points. The highlighted Ln means that line segments will be used to connect the points. Pnts means that the graph will be drawn as a set of points. Figure 10.21 shows the graph of $y =$ drawn with points only. The $\boxed{\texttt{/}}$ key acts as a toggle switch; each time you press $\boxed{\texttt{/}}$ the program switches back and forth between **/ ) Pnts/Ln** and **/ ) Pnts/Ln**.

**S) Resolutn.** This command allows you to set the resolution, or density of points to be plotted on your graphs. Larger values for resolution mean more points plotted; smaller values mean less points plotted. In fact, the value specified in the **S ) Resolutn** option <u>is</u> the number of points plotted. The speed of the plot is inversely related to the resolution. As the number of points increases, the speed of the plot decreases, and vice versa. The default resolution of 60 gives fairly accurate graphs, but often it is necessary to increase the resolution to obtain more accu-

rate graphs.    The results of changing the resolution will not be seen until the graph is replotted.

**P)   Print.** This command allows you to obtain a print-out of the current graph using some printers. Select
the **P) Print** command and the **Print Menu** will appear. Select the type of printer interface card
you have in your computer and press $\boxed{\text{Return}}$. The program will prompt you to prepare your printer;
press any key to print. When the printing is complete, the program will return to the **commands
menu**. Press $\boxed{\text{Esc}}$ to return to the **commands menu** without printing.

To produce a screen-dump of the entire screen including the graph and the **commands menu**, press
the $\boxed{\text{Control}}$ $\boxed{\text{P}}$ keys (at the same time). Follow the prompt at the bottom of the screen in the same
way listed above.

**Q)   Quit.** Select this command when you are ready to quit the program or to return to the **Main Menu**
to select another graphing program. After you select **Q ) Quit**, press any key to go to the **Main
Menu**. Press $\boxed{\text{Exc}}$ to abort the **Q ) Quit** command and return to the **commands menu**.

## 10.4 Apple II Examples

**10.4.1   Entering Functions:** Try graphing the following functions. A graph and a viewing rectangle
are given for each function. If you fail to obtain the same graph, check the keying sequence you used and
compare it with the ones given below for the four graphs.

(1)   $f(x) = x^2 - 3$                                    (2)   $f(x) = |x + 5|$

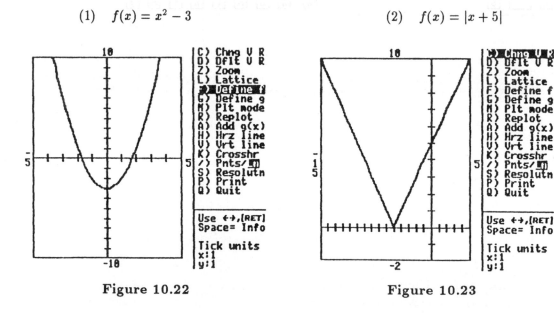

Figure 10.22                                    Figure 10.23

$$(3) \quad f(x) = (-1)(x-1)^2 \qquad\qquad\qquad (4) \quad f(x) = \frac{2}{x} + \frac{3}{(x-2)^2}$$

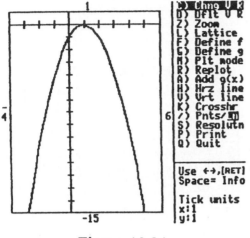

**Figure 10.24**

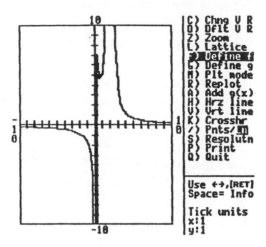

**Figure 10.25**

Were you able to produce these graphs? If not, here are the correct keying sequences for 1–4.

(1) $\boxed{X}\ \boxed{\wedge}\ \boxed{2}\ \boxed{-}\ \boxed{3}$ 
(2) $\boxed{A}\ \boxed{B}\ \boxed{S}\ \boxed{((}\ \boxed{X}\ \boxed{+}\ \boxed{5}\ \boxed{)}$

(3) $\boxed{((}\ \boxed{-}\ \boxed{1)}\ \boxed{*}\ \boxed{((}\ \boxed{x}\ \boxed{-}\ \boxed{1}\ \boxed{)}\ \boxed{\wedge}\ \boxed{2}$ 
(4) $\boxed{2}\ \boxed{/}\ \boxed{x}\ \boxed{+}\ \boxed{3}\ \boxed{/}\ \boxed{((}\ \boxed{x}\ \boxed{-}\ \boxed{2}\ \boxed{)}\ \boxed{\wedge}\ \boxed{2}$

**10.4.2  Zoom-out:** Zoom-out is another way to change the viewing rectangle. Let's look at an example. Change the function $f$ to $f(x) = x^2 - 23x + 132$. The graph of $f$ in the default viewing rectangle of $[-10, 10]$ by $[-10, 10]$ is given in Figure 10.26. Notice that this figure shows very little of the graph of $f$.

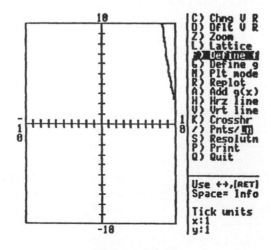

**Figure 10.26**

Select **O ) Zoom out** from the **Zoom menu** and zoom out horizontally (along the $x$-axis) by a factor of 3, and vertically (along the $y$-axis) by a factor of 4 to look at the graph in a larger viewing rectangle (see Figure 10.27). Try this on your own first. If you need some help, see the keying sequence below.

(1)  Select **Z ) Zoom** from the **commands menu**

(2)  Select **O ) Zoom out** from the **Zoom menu**

(3)  Answer the first prompt ("Please enter the zoom factor for the x axis:") with the keys ③ Return .

(4)  Answer the second prompt ("Please enter the zoom factor for the y axis:") with the keys ④ Return .

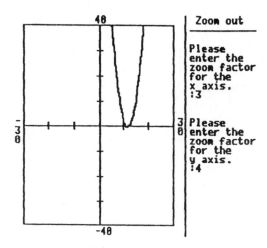

**Figure 10.27**

When you are not sure you have a complete graph of a function, you can use the **O ) Zoom out** command to quickly look for an appropriate viewing rectangle that displays a complete graph of the function. You often need to use a vertical zoom factor different from the horizontal zoom factor. For example, change the function $f$ to $f(x) = -5x^3 + 7x^2 + 6x + 25$. Zoom out using 10 for the zoom factor for both the horizontal and vertical direction. The new viewing rectangle will be $[-100, 100]$ by $[-100, 100]$. What happened? Select **D ) Dflt V R** to return to the default viewing rectangle. Zoom out again using a zoom factor of 1 for the $x$-axis (horizontal) and a factor of 100 for the $y$-axis (vertical).

**10.4.3  Zoom-in:** Change $f$ to the function $f(x) = -x^3 - 4x^2 + 3x + 5$. The graph of $f$ in the default viewing rectangle is shown in Figure 10.28. In this viewing rectangle the middle $x$-intercept appears to be very close to $-1$. Suppose we want a better approximation to this $x$-intercept. We can use the **I )** **Zoom in** command to obtain a closer view. We will use this command to form a box around this middle $x$-intercept; this box will be the new viewing rectangle. Select **I ) Zoom in** from the **Zoom menu**. Using the ← → ↓ ↑ keys, or the L R U D keys, move the crosshair to one corner of a box which will capture the selected $x$-intercept (see Figure 10.28). When the first corner is in place press Space Bar to set the corner. Next, use the ← → ↓ ↑ keys or the L R U D keys to stretch the box in such a way that it will capture the x-intercept under investigation (see Figure 10.29). When the box is set correctly, press Space Bar to draw the graph of the function in the new viewing rectangle.

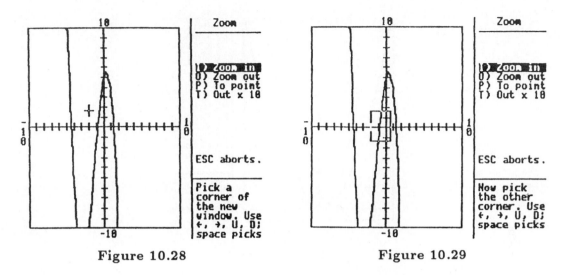

**Figure 10.28**                                            **Figure 10.29**

Figure 10.30 shows the graph of $f(x) = -x^3 - 4x^2 + 3x + 5$ drawn in the new viewing rectangle. Notice from this magnified view that the x-intercept we were investigating is actually in the interval $(-1, 0)$, and not at $-1$ as we suspected. You can continue to zoom in until you have located this $x$-intercept with as much accuracy as desired.

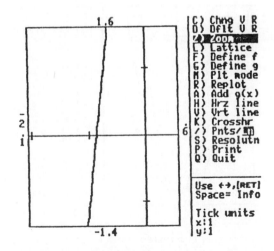

**Figure 10.30**

**10.4.4  Overlaying Additional Functions:** Select **F ) Define f** and enter $f(x) = x^2$ as the function $f$. Change the viewing rectangle using the **C ) Chng V R** command to $[-4, 4]$ by $[-1, 15]$. Select **G ) Define g** and enter $g(x) = 2x^2$ as the function $g$. Next, select **A ) Add g(x)** to overlay the graph of the function $g(x) = 2x^2$ on the graph of $f(x) = x^2$ (see Figure 10.31).

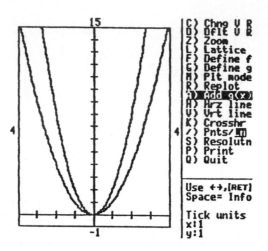

**Figure 10.31**

Next, overlay the graph of $g(x) = 4x^2$ to the viewing rectangle displaying the graphs of $f(x) = x^2$ and $g(x) = 2x^2$ (see Figure 10.32). This is done by using the **G ) Define g** and **A ) Add g(x)** commands again. What happens to the graph of $f(x) = ax^2$ as the coefficient, $a$, of $x^2$ gets larger?

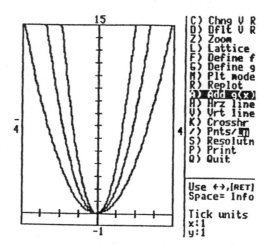

**Figure 10.32**

Select **R ) Replot** to draw only the graph of $f(x) = x^2$ in the present viewing rectangle. Change the viewing rectangle to $[-10, 10]$ by $[-1, 12]$ (see Figure 10.33). Using the **G ) Define g** and **A) Add g(x)** commands, add the following function to the viewing rectangle: $g(x) = \frac{1}{2}x^2$, $g(x) = \frac{1}{3}x^2$, and $g(x) = \frac{1}{10}x^2$. The keying sequence to enter the first function is ⟨ 1 / 2 ⟩ * x ^ 2 . The others are similar.

What can you say about the graph of $f(x) = ax^2$ as the coefficient of $x^2$ takes on fractional values between 0 and 1? Figure 10.34 shows these four functions overlaid on the same viewing rectangle.

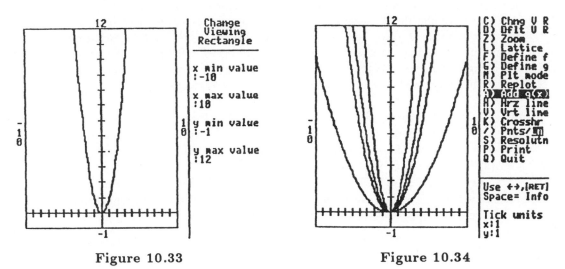

**Figure 10.33**                                    **Figure 10.34**

Select **D ) Dflt V R** to draw the graph of $f(x) = x^2$ in the default viewing rectangle. Overlay the graph of $g(x) = -x^2$ using the the keying sequence $\boxed{-}$ $\boxed{1}$ $\boxed{*}$ $\boxed{x}$ $\boxed{\char`\^}$ $\boxed{2}$ (see Figure 10.35). Notice we had to insert a $\boxed{1}$ in order to get the correct graph of $g(x) = -x^2$. What we entered then was $g(x) = -1x^2$. Your screen should look like the one in Figure 10.36. What can you say about the graph of $f(x) = ax^2$ if $a$ is negative?

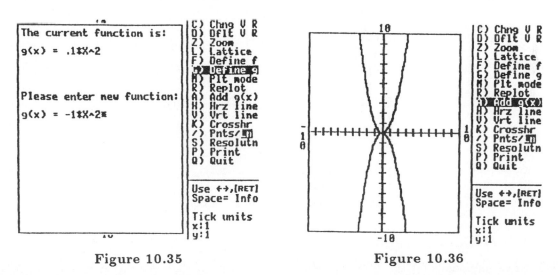

**Figure 10.35**                                    **Figure 10.36**

**10.4.5 Crosshair Lines:** Graph the function $f(x) = x^2 - 3x - 2$ in the default viewing rectangle. Select **K) Crosshr** from the **commands menu**. Use the $\boxed{\leftarrow}$ $\boxed{\rightarrow}$ $\boxed{\uparrow}$ $\boxed{\downarrow}$ keys or the $\boxed{L}$ $\boxed{R}$ $\boxed{U}$ $\boxed{D}$ keys to move the horizontal and vertical crosshair lines so that they intersect at some point of interest in the viewing rectangle. For example, use the crosshair to estimate the coordinates of the vertex of the parabola. Once the crosshairs are in position, press $\boxed{\text{Space Bar}}$ to see the coordinates of the point of interest.

To continue, press any key and you can move the crosshairs to another point in the viewing rectangle. To send the crosshairs to a specific location press $\boxed{\text{G}}$. A prompt in the lower right-hand corner of the screen will ask "Please type $x$ value to jump to:". Enter an $x$ value contained in the viewing rectangle and press $\boxed{\text{Return}}$. A second prompt asking "Please type $y$ value to jump to:" will appear. Enter a $y$ value contained in the viewing rectangle and press $\boxed{\text{Return}}$. You may continue this process as long as you wish. When you have finished using this command, press $\boxed{\text{Esc}}$ to return to the commands menu.

**10.4.6  Line and Point Plots:** Figure 10.37 shows the graph of $f(x) = \frac{3}{x+1}$ in the default viewing rectangle. There is something wrong with this graph. There appears to be a vertical line near the value $x = -1$. This line is **not** part of the graph of the function. It is called an **asymptote**. We can use the command **/) Pnts/Ln** to see that this line is not part of the graph.

Select **/ ) Pnts/Ln** and then **R ) Replot** to redraw the graph of the function in the **/ ) Pnts/Ln** mode. The function will be replotted using only points without the connecting line segments. Figure 10.38 shows the resulting screen display.

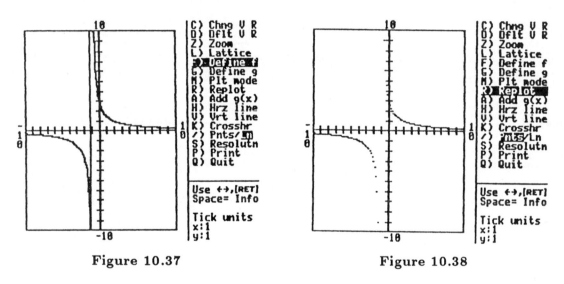

**Figure 10.37**                    **Figure 10.38**

Now we can see that the vertical line in Figure 10.37 is **not** part of the graph. You may select the **S ) Resolutn** command to increase the density of the points plotted in a graph. Drawing graphs with a computer is very fast, but you still have to be sure that you get a correct graph!

## 10.5  Apple II Conic Grapher

Most interactive commands for the *Conic Grapher* are the same as those for the *Function Grapher*. The default graph is $x^2 + y^2 - 9 = 0$, a circle with a radius of $3$ centered at the origin.

**E)  Defn Eqs.**  You will see this command on the **commands menu** in place of **F ) Define f** in the *Function Grapher*. This new command is used to enter conic equations and functions into the program. You may enter 2 different conic equations and 1 function in the program at once. Functions are entered in the same way as in *Function Grapher*. Conic equations must be entered in their general form

$$Ax^2 + Bxy + Cy^2 + Dx + Ey + F = 0$$

where $A, B, C, D, E$, and $F$ are the coefficients of the terms of the general conic equations. These 6 coefficients must be entered in order. For example, suppose you want to graph an ellipse with equation $\frac{x^2}{5^2} + \frac{y^2}{3^2} = 1$. First rewrite the equation in the general form $9x^2 + 25y^2 - 225 = 0$. Notice that $A = 9, B = 0, C = 25, D = 0, E = 0$, and $F = -225$. After each coefficient is entered, press $\boxed{\text{Return}}$. For coefficients which are zero, simply press $\boxed{\text{Return}}$ and the program will automatically enter the zero. Figure 10.39 shows the screen with the 6 coefficients for this ellipse entered and Figure 10.40 shows the graph of the ellipse.

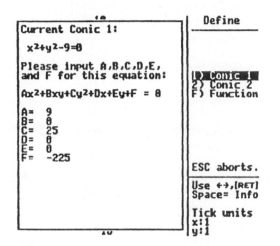

**Figure 10.39.** Enter the coefficients.

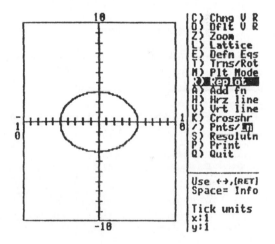

**Figure 10.40.** Graph of the ellipse.

The graph is an ellipse centered at the origin with major axis of length 10 on the $x$-axis and minor axis of length 6 on the $y$-axis.

The **E ) Defn Eqs** command is also used to enter the second conic equation or the function. The graph(s) which are automatically drawn on the screen depend on the **Plot Mode** settings chosen using the **M ) Plot mode** command. You may choose to draw the graph of Conic # 1, Conic # 2, or the function, or any combination of these three relationships.

**T)   Trns/Rot.** This command allows you to **translate** or **rotate** either conic equation you have defined. Translation of a conic graph is done by specifying horizontal and vertical translation factors for the center of the conic. For example, entering an $x$ translation factor of 3 and a $y$ translation factor of $-4$ for the ellipse defined above will redraw the graph of the ellipse centered at the point $(3, -4)$. Figure 10.41 shows this translated graph. Any further translations of the graph are always relative to the original graph and not to a translated graph.

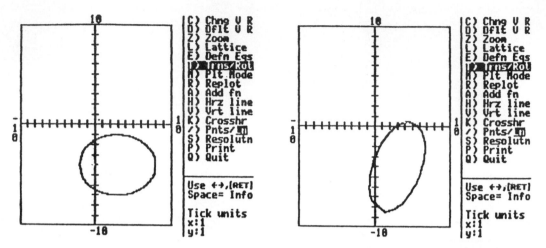

**Figure 10.41.** Translated ellipse.          **Figure 10.42.** Rotated ellipse.

This command also allows you to rotate any conic graph a specified number of **radians** about the origin. Positive values will rotate the graph counterclockwise and negative values will rotate the graph clockwise. As with translations, all rotations are relative to the original graph and not a rotated graph. However, once a graph is translated, rotation is done on the translated graph and vice versa. Figure 10.42 shows the graph of the translated ellipse defined above, rotated 1 radian counterclockwise.

**Space Bar Functions.** The $\boxed{\text{Space Bar}}$ has two important functions in the *Conic Grapher*. While a graph is being plotted, pressing the $\boxed{\text{Space Bar}}$ will pause the plotting process. Press any key to continue. At any other time, pressing the $\boxed{\text{Space Bar}}$ will display a window showing the conic equations and functions entered into the program, and the current translation and rotation factors.

All other commands in the *Conic Grapher* operate in the same way as those explained for the *Function Grapher*.

## 10.6  Apple II Parametric Grapher

Most interactive commands for the *Parametric Grapher* are the same as those for the *Function Grapher*. The default screen shows the graphs of two sets of parametric equations:

$$x_1 = 75 - 30T \quad y_1 = \sqrt{3}T = 16T^2$$
$$x_2 = 20\cos\tfrac{\pi T}{6} \quad y_2 = 20 + 20\sin\tfrac{\pi T}{6}.$$

The range of the parameter $T$ is $0 \leq T \leq 2$. Figure 10.43 shows the default screen; Figure 10.44 shows the equation information window activated by pressing the $\boxed{\text{Space Bar}}$. Pressing the $\boxed{\text{Space Bar}}$ while the graphs are plotting will pause the plot and print out the value of the parameter, $T$, and the coordinates of the last point(s) of the graph(s) plotted.

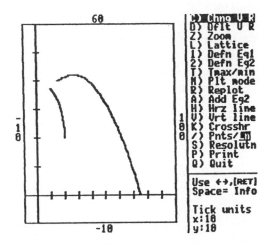

**Figure 10.43.** Default screen.

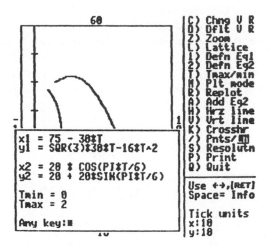

**Figure 10.44.** Equation information window.

**1 ) Defn Eq1** and **2 ) Defn Eq2.** These commands are used to enter the sets of parametric equations into the computer. Parametric equations are functions in terms of $T$ which are used to evaluate the $x$- and $y$-coordinates of each point. For example, select **1 ) Defn Eq1** and enter the parametric equations $x_1 = \sin(3T)$ and $y_1 = \cos(5T)$. After you enter the equations, the computer will ask for the range of $T$. Enter the values for **Tmin** and **Tmax**. For this problem enter **-PI** for **Tmin** and **PI** for **Tmax**. (The program recognizes the entry **PI** as $\pi$ and enters an 8-decimal-place approximation for the value.) Figure 10.45 shows the screen with the equations and the range of $T$ entered. Figure 10.46 shows the resulting graph in the viewing rectangle $[-1, 1]$ by $[-1, 1]$.

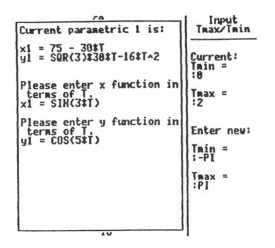

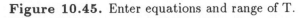

**Figure 10.45.** Enter equations and range of T.

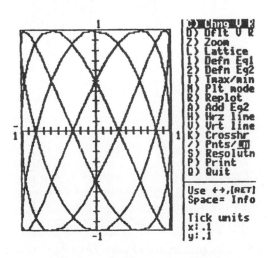

**Figure 10.46.** Resulting graph.

You may enter one or two sets of equations and, depending on the **Plot Mode** setting (set by using the **M) Plt mode** command), either one or both of the equations can be drawn automatically. **Make sure you enter equations in terms of $T$ and not $x$ as in the *Function Grapher*.**

**T) Tmax/min.** This command allows you to choose a new maximum and minimum value for the parameter $T$ of the function you are graphing. The procedure is identical to the last steps in the **1) Defn Eq1** command outlined above.

## 10.7 Apple II Polar Grapher

Most interactive commands for the *Polar Grapher* are the same as those for the *Function Grapher*. The default screen shows the graphs of $r(t) = 7\sin(3T)$ with the value of $T$ in the range $0 \le T \le \pi$.

**1) Defn Eq1** and **2) Defn Eq2.** These commands are used to enter the polar equations into the computer. Unlike the *Parametric Grapher*, each polar graph has only one equation. For example, the polar equation $r_1 = 10\cos(6T)$ is the only equation needed to define the graph (see Figure 10.47). Like the *Parametric Grapher*, you must enter a range for the parameter $T$ in terms of a **Tmin** and a **Tmax** value. Figure 10.48 shows the screen with the information window visible. The value of **Tmax** was entered as **2 * PI**. The computer understands **PI** to be $\pi$ and can do computations with **PI** within an input statement.

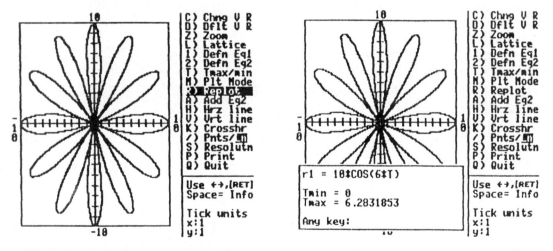

**Figure 10.47.** Graph of $r_1 = 10\cos(6T)$.     **Figure 10.48.** Information window visible.

**T) Tmax/min.** This command allows you to choose a new maximum and minimum value for the parameter, $T$, of the function you are graphing. This parameter represents the angular rotation from the positive $x$-axis in the polar coordinates $(r, \Theta)$. Often the range of $T$ will be in terms of $\pi$ or multiples of $\pi$. Use the keyboard equivalent of **PI** for $\pi$ or multiples of $\pi$ like **2*PI** or **PI/2**.

**K) Crosshr.** In the *Polar Grapher* you will notice a slight difference in the way the crosshairs work. When you have placed the crosshairs on a point whose coordinates you wish to see, press the ⌨Space Bar to see the $x$- and $y$-coordinates of the point in the lower right corner of the screen as usual. You will be prompted to press any key. When you press a key, the values for $T$ and $r$ will be displayed. Finally, pressing any key again will return you to the crosshair menu.

**S)** **Resolutn.** When graphing polar equations, it will often be necessary to increase the resolution of the graph to get accurate pictures. Resolutions of 120 or 180 will give better results than the default value of 60 .

*Important Notes*

(1) **PI** and **E** are acceptable constants for $\pi$ and $e$ .

(2) Any BASIC language expression may be input for <u>any</u> numerical input value. For example, you may enter **PI/2, SIN(3)**, or **ATN (E\*3)** in a "GOTO" or "Zoom factor" input statement.

(3) Pressing the [Esc] key gets you out of <u>anything</u>. If pressing [Esc] fails (it shouldn't ever), try pressing [Control] [Reset] at the same time. Pressing [Control] [Reset] returns to the **commands menu** and [Esc] returns to the active menu, such as the **Zoom menu** or the **Plot mode menu**.

## 10.8   Apple II 3-D Surface Grapher

**10.8.1   Introduction:** A great deal is known about single variable function graphers. The use of such graphers and understanding about how they can enhance the teaching and learning of mathematics are on the rise. Much less is known about the use of surface graphers, that is, devices that produce a graph of a function of two variables. However, several things about graphing functions of two variables are very clear. Obtaining graphs by hand is a difficult task for both students and teachers. Students have a good bit of trouble visualizing in three dimensions. Teachers have a difficult time producing quick, accurate graphs of functions of two variables.

The three dimensional grapher described in this guide is designed to allow the user to obtain reasonably accurate graphs of functions of two variables. The user can obtain the graph for $a \leq x \leq b$, $c \leq y \leq d$, and $e \leq z \leq f$, and then choose an arbitrary point in three dimensional space from which to view the graph. Once the first graph is drawn the points are stored in an array so that the graph can be redrawn quickly from different views. The user can choose any point in the three dimensional space from which to view the graph. The resolution of a graph is under user control.

The three dimensional grapher allows the user to interactively explore the behavior of surfaces. Local maximum and minimum values of the functions of two variables can be investigated graphically. The grapher can help students deepen their understanding and intuition about functions of two variables. It can provide a geometric representation of problem situations to go along with an algebraic representation. The connections between these two representations can be explored and exploited to gain better understanding about problem situations.

The single most important feature of this graphing program is that virtually every aspect of this utility is interactive and under user control. This utility was designed to help teachers teach and students learn mathematics in an atmosphere where both are active partners in the educational experience.

**10.8.2 Drawing and Viewing a 3-D Graph:** This section describes how the user chooses a region of three dimensional space in which to draw a graph of a function of two variables, and the way in which that graph can be viewed.

*Definition* The set $\{(x, y, z)|a \le x \le b, c \le y \le d, e \le z \le f\}$ is called the *viewing box* **[a, b] by [c, d] by [e, f]** where $a = x$-minimum, $b = x$-maximum, $c = y$-minimum, $d = y$-maximum, $e = z$-minimum, and $f = z$-maximum values of the viewing cylinder in the $xy$-plane.

Notice that the viewing cylinder $[a, b]$ by $[c, d]$ by $[e, f]$ is completely determined by the rectangular parallelepiped $a \le x \le b$, $c \le y \le d$, $e \le z \le f$ of the $xy$-plane. The user can change the viewing box by selecting the **C ) Chng V D** option from the **commands menu** and entering the new values corresponding to $x$-min, $x$-max, $y$-min, and $y$-max of the viewing cylinder. Figure 10.49 shows the default function, $f(x, y) = \sin(y)$, in the default viewing box of $[-10, 10]$ by $[-10, 10]$ by $[-10, 10]$. Figure 10.50 shows the screen for the **C ) Chng V D** option with the new viewing domain $[-4, 4]$ by $[-4, 4]$ by $[-4, 4]$ entered. Figure 10.51 shows the graph of $f(x, y) = \sin(y)$ redrawn in the new viewing cylinder $[-4, 4]$ by $[-4, 4]$ by $[-4, 4]$.

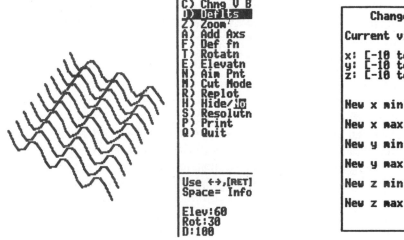

**Figure 10.49.** $f(x, y) = \sin(y)$.

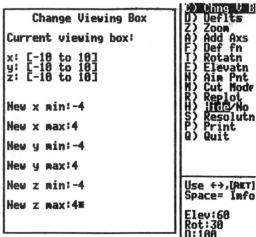

**Figure 10.50.** Change viewing domain screen.

Next, the user decides how to view the graph contained in the selected viewing cylinder. Two points can be selected. The point at which the user places his/her "eye" is called the *viewing point*. The point at which the view of the eye is directed is called the *aiming point*. The default aiming point is $(0, 0, 0)$, the origin of the three dimensional space. The aiming point can be changed by selecting **N) Aim Pnt** from the **commands menu** and inputting the rectangular coordinates of the point. The viewing point can be changed by entering the spherical coordinates $(d, \phi, \theta)$ of the point. Figure 10.52 shows the relationship between the aiming point, the viewing point, and the $xyz$- axes.

The spherical coordinate $d$ of the viewing point represents the distance from the aiming point to the viewing point. This parameter is controlled by selecting **Z ) Zoom** from the **commands menu**. The coordinate $\theta$ represents the angle that the plane perpendicular to the $xy$-axis containing the viewing point makes with the $x$-axis; the positive direction is counterclockwise. This parameter is controlled by selecting **T ) Rotatn** from the **commands menu**.

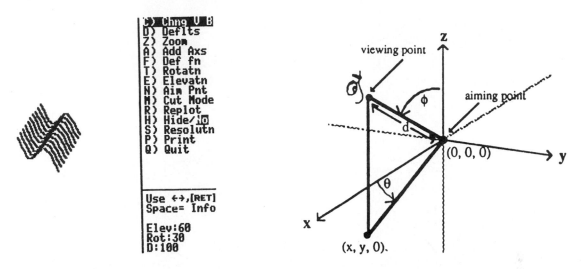

**Figure 10.51.** Graph of $f(x,y) = \sin(y)$ in $[-4,4]$ by $[-4,4]$ by $[-4,4]$.

**Figure 10.52.** Orientation in 3D space.

The coordinate $\phi$ represents the angle the line through the aiming point and the viewing point makes with the $z$-axis. A value of $\phi = 0°$ means you are looking directly down on the $z$-axis. A value of $\phi = 90°$ means you are looking perpendicular to the $z$-axis, parallel to the $xy$-plane. Value of $\phi$ between $90°$ and $180°$ means you are viewing the graph from below the $xy$-plane. This parameter is controlled by selecting **E ) Elevatn** from the **commands menu**. The default viewing point is $(100, 60°, 30°)$.

The software draws the graph in a cone of vision determined by the aiming point and the viewing point. The size of the cone is an automatic feature and cannot be selected by the user. The viewing point is the vertex of the cone, and the line determined by the viewing point and the aiming point is the axis of the cone. The view is from the viewing point toward the aiming point. The option **H) Hide/No** allows the user to view the graph with or without hidden lines. That is, if the hidden lines option is on, then the user will not see the portions of the surface that should be hidden from view by other portions of the surface. Basically, the graph that the user sees is the intersection of the software selected cone of vision with the user selected viewing cylinder. By changing the aiming point, viewing cylinder, and the viewing point, the user can view *any* portion of a surface with a high degree of resolution.

This drawing and viewing feature of the software literally allows the user to move around and view the surface as if in an airplane. You can move closer or further away, and view the graph above or below by careful selection of elevation, rotation, and distance (see Figure 10.53).

**10.8.3 Interactive Menu Commands for the 3-D Surface Grapher:** The following is a detailed explanation of the options on the **commands menu**.

**C) Chng V B** (Change viewing box). This command allows the user to change the dimensions of the viewing box. The default viewing cylinder is $[-10, 10]$ by $[-10, 10]$ by $[-10, 10]$. Figure 10.50 shows the screen with the new values $[-4, 4]$ by $[-4, 4]$ by $[-4, 4]$ entered. Once the values are entered, select **R ) Replot** from the **commands menu** to replot the graph in the new viewing cylinder.

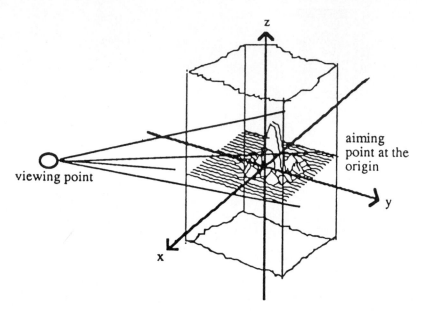

**Figure 10.53.** Orientation of the viewing point, aiming point and the graph.

**D) Deflts** (Defaults). This option replots the current graph with the default aiming point of $(0,0,0)$, the default viewing point of $(100, 60°, 30°)$, in the default viewing box of $[-10, 10]$ by $[-10, 10]$ by $[-10, 10]$, and using only $y$-axis cuts.

**Z) Zoom.** This option changes the distance from the viewing point to the aiming point. A distance of 100 (the default setting) causes the viewing box $[-10, 10]$ by $[-10, 10]$ to approximately "fill" the computer screen. Figure 10.54 shows the default function, $f(x,y) = \sin(y)$ drawn with a zoom factor of 200 and Figure 10.55 shows the same function drawn with a zoom factor of 50.

Changing the zoom factor does not change any of the other parameters of the graph relating to aiming point or viewing point. Select **R ) Replot** to redraw the graph.

**A) Add Axs** (Add Axes). This option allows the user to superimpose the $x$-, $y$-, and $z$-axes on the graph. Figure 10.56 shows the default graph of $f(x,y) = \sin(y)$ with the axes in place. To remove the axes, select **R ) Replot**.

**F) Def fn** (Define function). This option is used to enter a new function. Functions are entered in terms of $x$ and $y$ using standard AppleSoft BASIC language symbols. The constants $e$ and $\pi$ are entered as $\boxed{\text{E}}$ and $\boxed{\text{P}}$ $\boxed{\text{I}}$ respectively. Special keying instructions are needed to enter built-in functions or to obtain correct graphs of some special functions.

Here are some examples.

(1) The absolute value function $|x + y|$ is entered as **ABS ( X + Y)**.

(2) The natural logarithm function $\ln y$ is entered as **LOG ( Y )**. *Note* $y = \log_b(y)$ can be graphed for any base $b$ by entering **LOG ( Y ) / LOG ( B )**.

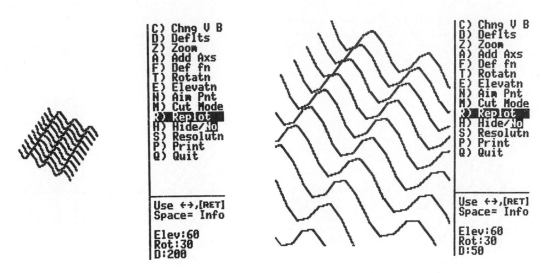

**Figure 10.54.** Zoom factor of 200.          **Figure 10.55.** Zoom factor of 50.

(3)   The exponential function $e^{x+y}$ is entered as **EXP ( X + Y )**

(4)   The root function $\sqrt[n]{x}$ for $n$ **odd** is entered as **ABS ( X ) / X * ABS ( X ) ^ ( 1 / N )**.

(5)   The greatest integer function $[x]$ is entered as **INT ( X )**.

(6)   The signum function is entered as **SGN ( Y )**.

(7)   The square root function is entered as **SQR ( X − Y )**.

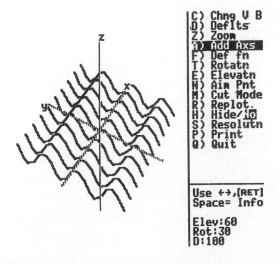

**Figure 10.56.** Axes superimposed on the graph.

**T)  Rotatn** (Rotation).  This option is used to change the angle, $\theta$ , that the plane perpendicular to the $xy$ -axis containing the viewing point makes with the $x$ -axis (see Figure 10.52). Positive values for $\theta$ indicate a counterclockwise rotation viewed from the top; negative values indicate a clockwise rotation. Figure 10.57 shows the default function at the default rotation of $\theta = 30°$ and Figure 10.58 show the same function rotated $\theta = 15°$ .

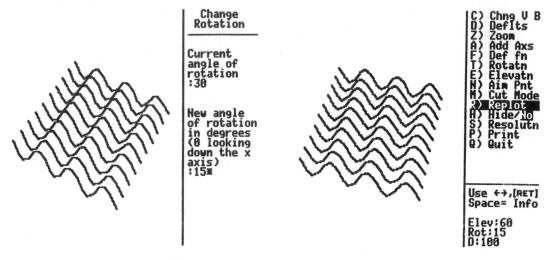

**Figure 10.57.** Default graph.            **Figure 10.58.** Rotation of $15°$ .

**E)  Elevatn** (Elevation).  This option controls the angle that the line from the viewing point to the aiming point makes with the $z$ -axis (see Figure 10.52). A value of $\phi = 0°$ means you are looking directly down the $z$ -axis from above. A value of $\phi = 90°$ means you are looking perpendicular to the $z$ -axis, parallel to the $xy$ -plane. Values of $0° \leq \phi \leq 90°$ mean you are viewing the graph from above the $xy$ -plane. Values of $90° \leq \phi \leq 180°$ mean you are viewing the graph from below the $xy$ -plane. The default value is $\phi = 60°$ . Figure 10.59 shows the default graph with an elevation of $\phi = 75°$ .

**N)  Aim Pnt** (Aim Point).  This option controls the location of the aiming point. The default aiming point is $(0, 0, 0)$ , the origin of the three dimensional space. The aiming point can be changed by inputting the rectangular coordinates of the point you wish to be the center of the viewing cone.

**M)  Cut Mode.**  Select this option to choose the axis through which the points of the graph will be plotted. The default cut mode is **y-cut only**. This means that the graph will be plotted based on values along the $y$ -axis. By selecting **2 ) y then x** from the **Cut Mode menu**, and **R) Replot** the graph will be drawn with both the $y$ - and $x$ -cuts (see Figure 10.60).

Using both $y$ - and $x$ -cuts gives a better representation of the 3D surface. However, using this drawing mode requires much more time for computation of points and drawing the graph.

**R)  Replot.**  Use this option to redraw the graph after changing any of the parameters.

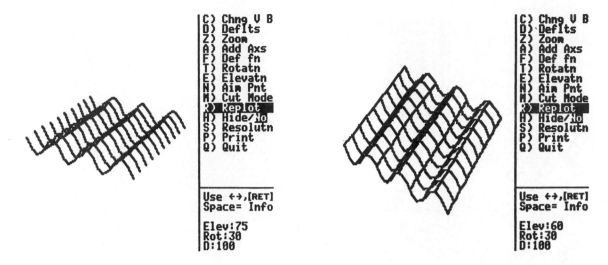

**Figure 10.59.** Elevation of $\phi = 75°$.       **Figure 10.60.** Cut Mode menu and default graph plotted with $y$- and $x$-cuts.

**H)** **Hide/No** (Hidden lines/ No hidden lines). This option allows the user to plot a graph with or without hidden lines. That is, if the hidden lines option is on, then the user will not see the portions of the surface that should be hidden from view by other portions of the surface. When the hidden lines option is off, all computed points will be plotted. Figure 10.61 shows the default graph with an elevation of $\phi = 75°$ plotted with the hidden lines option turned on. Notice that some of the surface is behind other portions giving a three dimensional look to the two dimensional representation.

**S)** **Resolutn** (Resolution). This option controls the number of sections and the number of points per section which are computed to draw the graph. The default graph is drawn with 10 $y$-cut sections of 20 points each. This means that $10 \times 20 = 200$ individual points are computed and then plotted on the screen. Increasing the number of sections and/or the number of points per section draws a graph with better resolution, but requires more time. For example, choosing the maximum number of sections (20) and the maximum number of points per section (70) in both the $x$ and $y$ direction will require the computation and plotting of 2800 individual points (as compared to 200 for the default graph). Figure 10.62 shows the default graph drawn with the maximum number of sections and points.

**P)** **Print.** This option prints the current graph to some printers.

**Q)** **Quit.** This option ends the current session and returns to the **Main Menu**.

### 10.8.4 Helpful Hints

(1) Press the $\boxed{\text{Esc}}$ key to abort the computation of points or the plotting of the graph. The $\boxed{\text{Esc}}$ key feature is useful when you select a high resolution and want to change it before waiting for the complete array of values to be computed or the graph to be drawn.

(2) The $\boxed{\text{Space Bar}}$ has two functions: 1) Pressing the $\boxed{\text{Space Bar}}$ while a graph is being plotted will cause the plot to pause and the coordinates of the last point plotted to appear in the window at the lower right of the screen. Press any key to continue the plot (see Figure 10.63). 2) Pressing the $\boxed{\text{Space Bar}}$

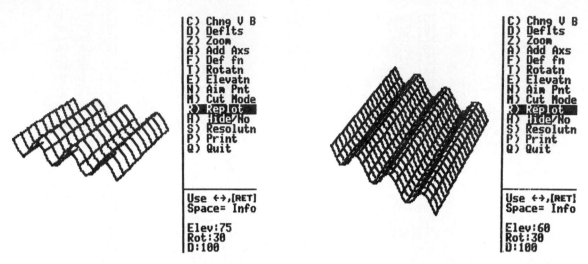

**Figure 10.61.** Hidden lines option turned on     **Figure 10.62.** Default graph drawn with maximum sections and points per section.

at any other time will show an information window listing the current function, elevation, rotation, distance, aiming point, and the number of sections and points per section used to draw the graph. Figure 10.64 shows the information window for the default function.

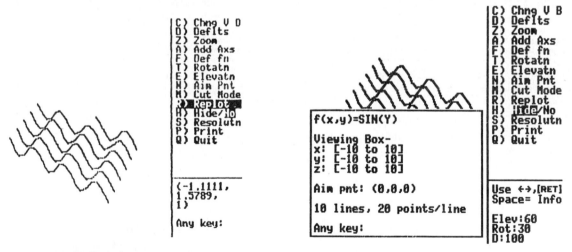

**Figure 10.63.** Graph paused.          **Figure 10.64.** Information window.

(3) Because of the increased plotting time, increase resolution only when you desire a "nice" plot.

(4) Use the **Z ) Zoom** option to change the cone of vision. Changing the distance $d$ allows you to see more, or less, of the graph contained in the viewing cylinder.

(5) To zoom-in on features away from the center (aiming point) of the current graph, you may need to change both the viewing cylinder and the aiming point. However, do this in low resolution until you are sure you have selected the viewing cylinder and aiming point that displays the desired features.

# Chapter 11

## Master Grapher Version 1.0 for the IBM<sup>R</sup>

### 11.1  *Master Grapher* Start Up

**IMPORTANT.**  Do NOT write protect either the *MASTER GRAPHER* or *3D-GRAPHER* disk; the program writes to the disk.

**Start Up.**  Boot the unit with DOS 2.1 (or greater) in drive A, then insert the *Master Grapher* disk into drive B. The program should run on a machine with only 256K of memory if it is run under DOS 2.X, however it will not run under DOS 3.X on the same machine because of the amount of memory required by DOS 3.X.

**Screen Dumps.**  Type *Graphics* after the A > prompt if you want to use the Shift PrtSc key for screen dumps of your graphs. To access drive B, type b : and press the return key. To run the graphing program, type *MASTER* and press the return key.

**Hercules Graphics Card.**  If you have a Hercules graphics card or compatible you will need to follow the following instructions instead of the preceding. To access drive B, type b : and press the return key. Type *PRTSCR* after the B > prompt if you want to use the Shift PrtSc key for screen dumps of your graphs. To run the graphing program, type *MASTER-H* and press the return key.

**Main Menu.**  Information about the graphing program will appear on the screen. Press any key to go to the *main menu*. The *main menu* will give a list of the four different graphing programs that are available for your use.

**3-D Grapher.**  *Master Grapher* includes *3-D Grapher*, a graphing utility for functions of *two* variables. The *3-D Grapher* utility is described in Sections 11.7–11.10.

### Main Menu

Type the appropriate key

(1)  Function Grapher

(2)  Conic Grapher

(3)  Parametric Grapher

(4)  Polar Grapher

(5)  Help

(6)  Exit

## 11.2 Function Grapher

When you are ready to enter the Function Grapher, select ☐1. There will be a slight delay while the function grapher program is loaded. The initial screen (except for the function graphed and this viewing rectangle) will look like the one shown in Figure 11.1.

The interactive commands available for your use will be in the menu on the right side of the screen under the Function Menu. In the lower right corner are listed the distance between horizontal scale marks (HS) and the distance between the vertical scale marks (VS), as well as the translation factors (XT and YT), and the rotation factors (R).

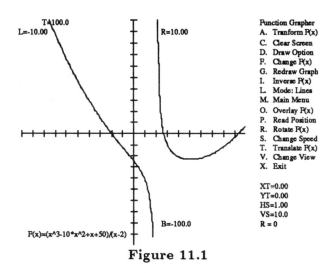

**Figure 11.1**

**11.2.1  Transform F(x):** This command allows you to draw the graph of $y = A f (B x + C) + D$, by specifying $f$, A, B, C, and D. Select ☐A; you will then be asked to enter the index of the function $f$ you wish to transform. Next you will be asked to enter the parameters A, B, C, D. You may enter any real number or algebraic expression (for example, "pi*2"). The function $y = A f (B x + C) + D$ will be plotted immediately.

**11.2.2  Clear Screen:** Select ☐C to clear the graphing screen.

**11.2.3  Draw Option:** Select ☐D. A menu like the one to the right will appear with a list of options. Use ☐1 through ☐4 to approximate the coordinates of a point on a graph. To approximate the $y$-coordinate, a horizontal line can be specified or moved up and down (use the ☐↑ or ☐U to move the line up, the ☐↓ or ☐D to move the line down, and ☐S to change the rate of movement of the line) until it is fixed in position by pressing any other key. The $y$-coordinate of each point on the horizontal line will be displayed. The $x$-coordinate can be found in a like manner using a vertical line (use the ☐→ or ☐R and the ☐← or ☐L to move it right and left). Select ☐5 to overlay a lattice (an array of dots) in the viewing rectangle. The *distance* between these dots will be given by the HS (horizontal scale) and VS (vertical scale) values found in the lower right-hand corner of the screen. This command will be helpful to estimate the coordinates of a point on a graph. For example, zoom in on some area of the current graph that is of interest to you. Select

$\boxed{\text{D}}$ $\boxed{\text{5}}$ to obtain a lattice in the viewing rectangle. Use the HS and VS values to read the coordinates of the point you selected.

**Draw Option**

(1)   Moving V. Line

(2)   Specified V. Line

(3)   Moving H. Line

(4)   Specified H. Line

(5)   Grid

(6)   Previous Menu

**11.2.4   Change F(x):** Select $\boxed{\text{F}}$ to change the function to be graphed. You will see a menu with 10 options. Pressing $\boxed{\text{1}}$ through $\boxed{\text{8}}$ will toggle between *Not Displayed* and *Displayed*. Only the *Displayed* function(s) will be plotted. To change a function from Displayed to Not Displayed (or vice versa) type the index of the function whose status you wish to change. Thus, you can *predefine* and plot up to 8 functions in this manner. Select $\boxed{\text{9}}$ to return to the default menu.

Select $\boxed{\text{0}}$ to change any one of the eight functions listed. When you press $\boxed{\text{0}}$, you will then be asked which function index you wish to change. (There will be no prompt). You will now be able to edit the function you specified; to do this you can use keys in the table below. Edit the equation you want to graph (in terms of $x$ in the function and conic grapher and in terms of $t$ in the polar and parametric graphers), then press $\boxed{\texttt{return}}$.

**Editing Keys**

| | |
|---|---|
| $\boxed{\rightarrow}$ | Moves the cursor to the right one character. If you are at the end of the equation this has no effect. |
| $\boxed{\leftarrow}$ | Moves the cursor to the left one character. If you are at the beginning of the equation this has no effect. |
| $\boxed{\texttt{backspace}}$ | Has the same effect as the $\boxed{\leftarrow}$ key. |
| $\boxed{\texttt{Del}}$ | Deletes the character at the current cursor position. |
| $\boxed{\texttt{End}}$ | Moves the cursor to the end of the equation. |
| $\boxed{\texttt{Esc}}$ | Deletes the edited equation and sets the cursor to the first position. |
| $\boxed{\texttt{Home}}$ | Moves the cursor to the end of the equation. |
| $\boxed{\texttt{Ins}}$ | Changes insert mode from overstrike to insert. Pressing it again reverses the process. |
| $\boxed{\texttt{Return}}$ | Truncates the equation where the cursor is and exits edit mode. If an error is made in the equation it will revert back to the original equation. |

Use standard BASIC syntax to enter the desired equation. To enter $x^4 - 3x^2 + 15$ for example you would enter $\boxed{\text{x}}$ $\boxed{\wedge}$ $\boxed{\text{4}}$ $\boxed{-}$ $\boxed{\text{3}}$ $\boxed{*}$ $\boxed{\text{x}}$ $\boxed{\wedge}$ $\boxed{\text{2}}$ $\boxed{+}$ $\boxed{\text{1}}$ $\boxed{\text{5}}$. The constants $e$ and $\pi$ are entered as $\boxed{\text{e}}$ and $\boxed{\text{p}}\boxed{\text{i}}$, respectively. Special keying instructions are needed to enter built-in functions or to obtain correct graphs of some special functions. Here is a list of the special functions.

## Special Symbols and Built-in Functions

( 1)  $+$  is addition.                ( 2)  $-$  is subtraction.              ( 3)  $*$  is multiplication.
( 4)  $/$  is division.                ( 5)  $\wedge$  is  $x^a$ .            ( 6)  $\backslash$  is Integer Division.
( 7)  ABS$( x - 2 )$  is  $| x - 2 |$ .      ( 8)  CEIL$( x ) = [x + 1]$ .
( 9)  EXP$( x )$  is  $e^x$ .          (10)  FIX$( x )$  is FLOOR if  $x > 0$ , and CEIL if  $x < 0$ .
(11)  FLOOR$( x ) = [\![x]\!]$ .        (12)  INT$( x )$  is the greatest integer function .
(13)  LOG$( x )$  is  $\ln x$ .          (14)  LOG10$( x )$  is  $\log_{10}(x)$ .
(15)  LOG2$( x )$  is  $\log_2(x)$ .      (16)  ROUND$( x )$  rounds to the nearest integer.
(17)  SGN$( x )$  is the signum function.    (18)  SQR$( x + 6 )$  is  $\sqrt{x + 6}$ .
(19)  SIN$( x )$  is  $\sin x$ . All other trigonometric functions are entered in the same manner (e.g. arctan  $x$  is ARCTAN$( x )$ , cosh  $x$  is COSH$( x )$ , etc.). Here is a list of all the trigonometric functions supported: arccos, arccosh, arccot, arccoth, arccsc, arccsch, arcsec, arcsech, arcsin, arcsinh, arctan, arctanh, cos, cosh, cot, coth, csc, csch, sec, sech, sin, sinh, tan, tanh.

## Special Functions

(20)  LOGB$( x, a )$  is  $\log_a(x)$ .
(21)  ROOT$( x, a )$  is  $x^{1/a}$ . Note: you can enter  $x^{1/a}$  as  $x \wedge (1/a)$ , but you will only get the portion of the graph in the first quadrant.
(22)  POWER$( x, a )$  is  $x^a$ .

**11.2.5  Redraw Graph:** Redraws the graph with the current settings.

**11.2.6  Inverse F(x):** If you wish to view the inverse relation  $(y, x)$  where  $y = f(x)$  of any function in the function menu, select  $\boxed{\text{I}}$ . Then select the index of the function you wish to invert and press  $\boxed{\text{return}}$ , and the inverse *relation* will be overlayed immediately.

**11.2.7  Mode: Lines or Points:** Select  $\boxed{\text{L}}$  to choose one of two plotting modes. One plots only the points evaluated for a particular function, the second connects each consecutive pair of points with a line segment. This is the default mode.

**11.2.8  Main Menu:** Select  $\boxed{\text{M}}$  to return to the main menu.

**11.2.9  Overlay F(x):** When you select  $\boxed{\text{O}}$  you will need to select an index function to overlay and then press  $\boxed{\text{return}}$ . For example, if you choose 6, then the sixth function in the function menu will be plotted on the same screen with the function(s) already plotted. Once the index is selected, you will be asked whether or not if you want the function rotated and translated. If there are non-zero values in the XT, YT, and R categories in the lower right hand corner of the display screen, then choosing  $\boxed{\text{1}}$  and pressing  $\boxed{\text{return}}$  will have the selected function plotted *with* those rotation and translations applied. Choosing  $\boxed{\text{O}}$  will have the function plotted *without* any rotation or translation.

**11.2.10  Read Position:** The next option  $\boxed{\text{P}}$  can also be used to approximate the coordinates of any point in the current viewing rectangle. When you select  $\boxed{\text{P}}$ , a cursor will appear in the middle of the viewing rectangle. This cursor can be moved by using the arrow keys or pressing  $\boxed{\text{U}}$ ,  $\boxed{\text{D}}$ ,  $\boxed{\text{L}}$ ,  $\boxed{\text{R}}$ . To change the rate of movement of the cursor, type  $\boxed{\text{S}}$ . The speed control cycles from slow to fast so you may have to experiment with the speed each time you use it. When the cursor is at the desired location, type any other key. In the lower right-hand corner you will now see the location of the cursor in the viewing rectangle.

**11.2.11  Rotate F(x):**  Select $\boxed{\text{R}}$ to rotate the function about the origin. When you select $\boxed{\text{R}}$, you will need to enter the counterclockwise rotation angle in degrees. This can be either a real number or an algebraic expression like "2*180/pi". A positive input will produce a counterclockwise rotation and a negative input will be clockwise.

**11.2.12  Change Speed:**  To change the plotting speed, press $\boxed{\text{S}}$. You will now see the current plotting speed, the minimum and maximum allowable plotting speeds, and a prompt to enter a new speed. Enter the speed you want by typing any real number or algebraic expression with terms pi, e, or some other constants and then press $\boxed{\texttt{return}}$. The program will evaluate the expression and enter the resulting value as the new plotting speed—the larger the number of points evaluated, the slower the plot and the better the resolution. (You may wish to experiment with various settings until you find the one you prefer.) The default speed is a good compromise between speed and resolution.

**11.2.13  Translate F(x):**  Select $\boxed{\text{T}}$ to draw a graph of the function $y = f(x)$ translated H units horizontally and V units vertically. After selecting $\boxed{\text{T}}$, enter the amount of horizontal translation and press $\boxed{\texttt{return}}$. Then enter the amount of vertical translation and press $\boxed{\texttt{return}}$. Nothing will appear to happen. To see the effect of the translation, you will need to redraw the graph $\boxed{\text{G}}$. You will now have graphed the *displayed* function with the translation factors applied. If you wish to view both the original function and the translated function, select $\boxed{\text{O}}$ Overlay F(x) and choose the index of the original function and then $\boxed{\text{O}}$ W/O Rot. & Trans. In this manner you can get both the function and its translation in the same viewing rectangle.

**11.2.14  Change View:**  Selecting $\boxed{\text{V}}$ will display the menu below. You may have changed the speed, the plot mode or a variety of different commands that do not cause the graph to be immediately replotted; the function grapher incorporates those changes when it is redrawn or the viewing rectangle is changed. Suppose we want to view the graph of

$$f(x) = \frac{x^3 - 10x^2 + x + 50}{x - 2}$$

in the viewing rectangle $[-1, 1]$ by $[-1, 1]$. To do this enter $f$ by selecting $\boxed{\text{F}}$ and proceeding as detailed in the "Change F(x)" section (11.2.4), then select $\boxed{\text{V}}$. You can now choose the method you want to change the viewing rectangle.

**View Menu**

(0)   Zoom In

(1)   Zoom In (Point)

(2)   Zoom Out

(3)   Zoom Out (Point)

(4)   Set Zoom Factor

(5)   Set Window

(6)   Default Window

(7)   Last Window

(8)   X-Scale: 1

(9)   Previous Menu

(0)   **Zoom-In.**  To set your own area for the new viewing rectangle, select ⬚0. Place the pointer at the desired corner of the zoom–in rectangle, then press the space bar to set the corner. Now use the arrow keys to draw the desired viewing rectangle on the screen. Press the space bar to obtain a new plot. The arrow keys are on the numeric key pad. To move left, type ⬚← or ⬚L; to move right, type ⬚→ or ⬚R; to move up, type ⬚↑ or ⬚U; to move down, type ⬚↓ or ⬚D. This time you will be moving the cursor around to locate a corner of the new viewing rectangle. Position the cursor as you did the last time. Now, use **only** the arrow keys to move toward the corner opposite the one you just marked for the new viewing rectangle. As you move in one direction, a straight line forms. Moving perpendicular to the initial direction you chose will cause a rectangle to form. The one corner you selected initially will remain fixed and you will be moving the opposite corner around. When you are satisfied with the location of the opposite corner of the box, press the space bar. You will now see the function plotted in the new viewing rectangle.

An illustration of the effect that "zoom-in" can have on a given view appears in Figure 11.2. The function $f(x) = x^2 - 3x + 2$ was graphed in the standard viewing rectangle, then graphed again in the $[0.8, 2.3]$ by $[-1, 0.5]$ viewing rectangle. The behavior of the graph between $x = 1$ and $x = 2$ is much more obvious in the zoom-in view.

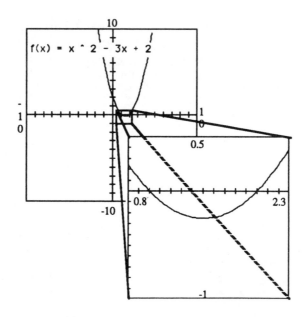

**Figure 11.2**

(1) **Zoom In (Point).** This command is useful when you want a close-up or magnified view of a particular section of a graph around a particular point. Select $\boxed{1}$. Use the arrow keys or U, D, L, R to move the cursor. Type "S" to change the speed of the cursor's movement. Type any other key to set the position. Once again you will use the arrow keys or the U, D, L, and R keys to move the cursor around on the screen. Use these keys to move the cursor in the current viewing rectangle to a point that you want to be the center of a new viewing rectangle. (*Note* Initially the cursor will be centered on the origin.) When you are satisfied with a location, press any key. The function will be replotted in a new viewing rectangle with the point you selected at the center of the new viewing rectangle. The *size* of the viewing rectangle will be determined by the zoom factor settings (key 4).

(2) **Zoom Out.** When you select $\boxed{2}$, the graph will be immediately redrawn in a viewing rectangle expanded by the zoom factors you set with key $\boxed{4}$ (or the default values of 10 for $x$ and 10 for $y$).

(3) **Zoom Out (Point).** When you select $\boxed{3}$, you will be asked to move the cursor to the location about which you wish the "zoom-out" to be centered in exactly the way explained above in *Key 1, Zoom In (Point).* Once again the zoom factor will be determined by the default factors (both 10) or the factors you set using key $\boxed{4}$.

(4) **Set Zoom Factors.** This command sets the horizontal ($x$) and vertical ($y$) zoom factors. When you select $\boxed{4}$ you will see the old setting for $x$ and $y$ zoom factors and be prompted to enter new ones. This can be done by typing in any real number or algebraic expression. Assume the current viewing rectangle is [L, R] by [B, T]. Enter the value you wish [L, R] and [B, T] to be multiplied by. Values greater than one cause the horizontal size of the rectangle to increase (zoom-out), and values less than one cause the horizontal or vertical size of the rectangle to decrease (zoom-in). The change in size of the viewing rectangle is symmetric about the center of the rectangle. Sometimes you will want the zoom factors on the $x$-axis and $y$-axis to be different. Note: You can "zoom-in" using "zoom-out" by selecting zoom factors that are less than one.

(5) **Set Window.** To change the viewing rectangle, press $\boxed{5}$. The message screen will clear and you will see the current viewing rectangle settings where L, R, B, T are respectively left, right, bottom, and top of the viewing rectangle. It will now prompt you to enter the new parameters, starting with L. Type the real number or algebraic expression you prefer and press $\boxed{\text{return}}$. Next, you will enter the R, B, and T values. Answer each prompt with the desired real number or algebraic expression and press $\boxed{\text{return}}$. When $\boxed{\text{return}}$ is pressed after entering T, the screen will be redrawn in the specified viewing rectangle and the program will return to the "View Menu".

(6) **Default Window.** When you select $\boxed{6}$, the displayed function(s) will be replotted in the default viewing rectangle ([$-10, 10$] by [$-10, 10$]), speed (100 for the function grapher, 50 for the conic grapher and parametric grapher, and 200 for the polar grapher), mode (lines), rotation ($0°$), and translation $(1, 1)$. After the screen is redrawn you will be returned to the "View Menu".

(7) **Last Window.** When you select $\boxed{7}$, the displayed function(s) will be replotted in the last viewing rectangle. However speed, mode, rotation, and translation will remain unaltered from the current settings. After the screen is redrawn you will be returned to the "View Menu".

(8) **X-Scale.** When you select $\boxed{8}$, the distance of the tick marks will be changed on the $x$-axis. If X-Scale is in units of $\frac{\pi}{2}$ the distance between tick marks will be in a factor of units of either $\frac{\pi}{2}$ or $\frac{\pi}{4}$. If X-Scale is in units of 1 the distance between tick marks will be in a factor of units of either 1 or 0.5.

The graph will be redrawn immediately with the appropriate $x$-scale. Note: This option is available only in the function grapher.

(9) **Previous Menu.** When you select $\boxed{9}$, this will return you to the grapher menu. Note: This option is $\boxed{8}$ on all the other graphers.

**11.2.15  Exit:** Select $\boxed{X}$ to return to the DOS operating system.

## 11.3  Conic Grapher

When you are ready to enter the Conic Grapher, select $\boxed{2}$ from the Main Menu. There will be a slight delay while the conic grapher program is loaded. The initial screen (except for the conic equation graphed) will look like the one shown in Figure 11.3.

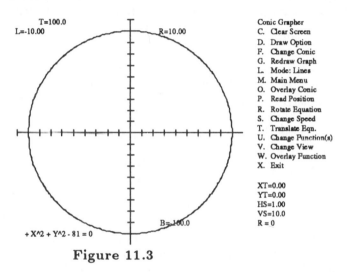

**Figure 11.3**

Notice the changes in the conic grapher menu versus the function grapher menu. The "A. Transform F(x)" and "I. Inverse F(x)" options have been removed. The "F. Change F(x)" is now "F. Change Conic" and the "O. Overlay F(x)" is now "O. Overlay Conic". Notice the "F(x)" in rotate and translate has been changed to equation. Finally, two new options have been added: "U. Change Function(s)" and "W. Overlay Function(s)". In the following sections we will only discuss these features as all the others have not changed. Note: Rotate and translate have not been changed; only the wording in the menu has been changed.

**11.3.1  Change Conic:** Pressing $\boxed{F}$ will clear the viewing rectangle and you will see a list of 8 *conic equations*. Selecting any of 1–8 will allow you to toggle between *Displayed* and *Not Displayed*. There must always be at least *one* conic equation displayed. Selecting $\boxed{9}$ will take you back to the conic grapher menu. Selecting $\boxed{0}$ will allow you to change a conic equation. When you press $\boxed{0}$ you will be prompted to enter the conic index you wish to change. After entering the index you will be prompted to enter the parameters A, B, C, D, E, and F in that order. These parameters are the constants in the following equation: $Ax^2 + Bxy + Cy^2 + Dx + Ey + F = 0$. The constants may be entered as any real number or algebraic expression.

**11.3.2  Overlay Conic:** Pressing $\boxed{\text{O}}$ will allow you to overlay a *conic equation* in the same manner you did in Section 11.2.9.

**11.3.3  Change Function(s):** Pressing $\boxed{\text{U}}$ will allow you to change or display any of the eight *functions* in the same manner you did in Section 11.2.4.

**11.3.4  Overlay Function:** Pressing $\boxed{\text{W}}$ will allow you to overlay one of the eight *functions* in the same manner you did in Section 11.2.9.

## 11.4  Parametric Grapher

When you are ready to enter the parametric grapher, select $\boxed{3}$ from the Main Menu. There will be a slight delay while the parametric grapher program is loaded. The initial screen (except for the parametric equation graphed) will look like the one shown in Figure 11.4.

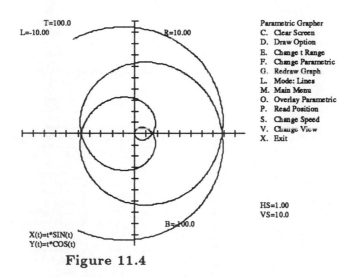

**Figure 11.4**

Notice the changes in the parametric grapher menu versus the function grapher menu. The "Transform F(x)", "Inverse F(x)", "Rotate Equation" , and "Translate Eqn." options have been removed. The "F. Change F(x)" is now "F. Change Parametric" and the "O. Overlay F(x)" is now "O. Overlay Parametric". Finally, a new option has been added: "E. Change t Range". In the following sections we will only discuss these features as all the others have not changed. This grapher will not plot the equations in the order of appearance in the "Change Equation" menu, but instead it plots the equations simultaneously so you can see the graphs drawn for the same $t$ .

**11.4.1  Change Parametric:** Select $\boxed{\text{F}}$ to change the parametric equation to be graphed. You will see a menu with six options. Pressing $\boxed{1}$ through $\boxed{4}$ will toggle between *Not Displayed* and *Displayed*. Only the *Displayed* parametric equation(s) will be plotted. To change a parametric equation from Displayed to Not Displayed (or vice versa), type the index of the function whose status you wish to change. Thus, you can *predefine* and plot up to four parametric equations in this manner. Select 5 to return to the default menu.

Select 0 to change any one of the four parametric equations listed. When you press 0, you will then be asked which function index you wish to change (there will be no prompt). You will now be able to edit the function you specified. To do this you can use the editing keys described in Section 11.2.4. After entering the index you will be prompted to edit the first part of the equation X(t) (in terms of *t* ). Press `return` when done. Now you will be prompted to edit the second part of the equation Y(t) (in terms of *t* ). Press `return` when done.

**11.4.2  Overlay Parametric:** Pressing 0 will allow you to overlay a *parametric equation* in the same manner you did in Section 11.2.9. Note that there are only four indices not eight as in all the other graphers.

**11.4.3  Change t Range:** Select E to change the bounds of *t* . The default bounds on the parameter *t* are $-10 < t < 10$ . After pressing E, you will see the current range of *t* , and then you will be prompted to enter tmin and tmax. These parameters can be any real number or an algebraic expression like "pi*2".

## 11.5  Polar Grapher

When you are ready to enter the polar grapher, select 4 from the Main Menu. There will be a slight delay while the polar grapher program is loaded. The initial screen (except for the polar equation graphed) will look like the one shown in Figure 11.5.

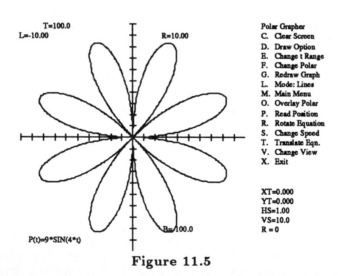

**Figure 11.5**

Notice the changes in the polar grapher menu versus the function grapher menu. The "A. Transform F(x)" and "I. Inverse F(x)" options have been removed. The "F. Change F(x)" is now "F. Change Polar" and the "O. Overlay F(x)" is now "O. Overlay Polar". Notice the "F(x)" in rotate and translate has been changed to equation. Finally, notice the new feature "E. Change t Range". In the following sections we will only discuss these features as all the others have not changed. Note that rotate and translate have not been changed; only the wording in the menu has been changed.

**11.5.1  Change Polar:** Pressing $\boxed{\text{F}}$ will allow you to change or display any of the eight *polar equations* in the same manner as you did in Section 11.2.4.

**11.5.2  Overlay Polar:** Pressing $\boxed{\text{O}}$ will allow you to overlay a *polar equation* in the same manner as you did in Section 11.2.9.

**11.5.3  Change t Range:** Pressing $\boxed{\text{E}}$ will allow you to change the t range in the same manner as you did in Section 11.4.3.

## 11.6  Special Keys

This section will discuss special keys available to the user.

(1)  **Print Graphics Screen.** If you have installed "graphics" you can use the $\boxed{\text{shift}}$ $\boxed{\text{PrtSc}}$ keys to get a screen dump.

(2)  **Abort Plot.** To STOP the plot at any time during the plotting routine and return to the menu, press the $\boxed{\text{Esc}}$ key or $\boxed{\text{E}}$ key.

(3)  **Pause Plot.** To interrupt (PAUSE) the plot, press the $\boxed{\text{spacebar}}$. Pressing the spacebar again or any other key will RESUME the plot.

## 11.7  3-D Grapher Introduction

A great deal is known about single variable function graphers. The use of such graphers and understanding about how they can enhance the teaching and learning of mathematics are on the rise. Much less is known about the use of surface graphers, that is, devices that produce a graph of a function of two variables. However, several things about graphing functions of two variables are very clear. Obtaining graphs by hand is a difficult task for both student and teacher. Students have a good bit of trouble visualizing in three dimensions. Teachers have a rough time producing quick, accurate graphs of functions of two variables.

The three-dimensional grapher described in this manual is designed to allow the user to obtain reasonably accurate graphs of functions of two variables. The user can obtain the graph for $a \leq x \leq b$, $c \leq y \leq d$ and $e \leq z \leq f$, and then choose an arbitrary point in three dimensional space from which to view the graph. Once the first graph is drawn, the points are stored in an array so that the graph can be redrawn quickly from different views. The user can choose any point in three-dimensional space from which to view the graph. The resolution of a graph is under user control.

This three-dimensional grapher allows the user to interactively explore the behavior of surfaces. Local maximum and minimum values of functions of two variables can be investigated graphically. The grapher can help students deepen understanding and intuition about functions of two variables. It can provide a geometric representation of problem situations to go along with an algebraic representation. The connections between these two representations can be explored and exploited to gain better understanding about problem situations.

The single most important feature of this graphing program is that virtually every aspect of this utility is interactive and under user control. This utility was designed to help teachers teach and students learn mathematics in an atmosphere where both are active partners in the educational experience.

We will now describe how the user chooses a region of three-dimensional space in which to draw a graph of a function of two variables, and the way in which that graph can be viewed.

*Definition* The set $\{(x, y, z) \mid a \leq x \leq b,\ c \leq y \leq d,\ e \leq z \leq f\}$ is called the *viewing box* $[a, b]$ by $[c, d]$ by $[e, f]$.

Notice that the viewing box $[a, b]$ by $[c, d]$ by $[e, f]$ is completely determined by the points $a \leq x \leq b$, $c \leq y \leq d$, $e \leq z \leq f$. The user can change the viewing box by selecting key [V] and entering the values of $a, b, c, d, e$, and $f$. The default viewing box is $[-10, 10]$ by $[-10, 10]$ by $[-10, 10]$.

Next, the user decides how to view the graph contained in the selected viewing box. Two points can be selected. The point at which the user places his/her "eye" is called the *viewing point*. The point at which the view of the eye is directed is called the *aiming point*. The aiming point can be changed by selecting key [A] and inputting the rectangular coordinates of the point. The viewing point can be changed by entering the spherical coordinates $(d, \phi, \theta)$ of the point using keys [Z], [R], and [E]. The "Change Elevation" key [E] allows the user to select the angle $\phi$. $\phi$ is the angle the line through the origin and the viewing point makes with the $z$-axis (Figure 11.6). The "Change Rotation" key [R] allows the user to select the angle $\theta$. $\theta$ is the angle between the $x$-axis and the plane perpendicular to the $xy$-plane which contains the viewing point (Figure 11.6). Positive direction is counterclockwise. Finally, the "Chg. Dist. (Zoom)" key [Z] allows the user to select the distance that the viewing point is from the aiming point. The default aiming point is $(0, 0, 0)$ and the default viewing point is $(100, 60°, 30°)$.

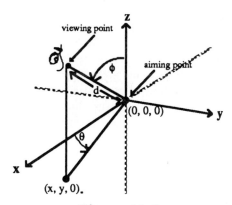

**Figure 11.6**

The software draws the graph, in a cone of vision determined by the aiming point and the viewing point. The size of the cone is an automatic feature and cannot be selected by the user. The viewing point is the vertex of the cone, and the line determined by the two points is the axis of the cone. The view is from the viewing point toward the aiming point. Selecting [H] allows the user to view the graph with or without hidden lines. That is, if the *hidden lines* option is on, then the user will *not* see the portions of the surface that should be hidden from view by other portions of the surface. Basically, the graph that the user sees is the intersection of the software-selected cone of vision with the user-selected viewing box. By changing the aiming point, viewing box, and viewing point, the user can view *any* portion of a surface with a high degree of resolution.

This drawing and viewing feature of the software literally allows the user to move around and view the surface as if in an airplane. You can move closer or farther away, and view the graph above or below by careful selection of elevation, rotation, and distance (Figure 11.7).

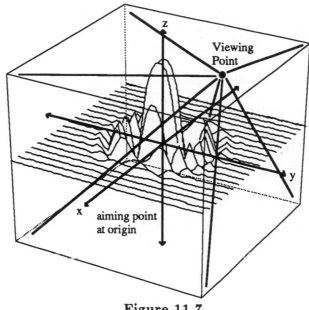

**Figure 11.7**

## 11.8  3-D Grapher Start Up

**Start Up.** Exit the *Master Grapher* program by pressing ⊠ and then do the following. If you are using the $5\frac{1}{4}''$ disk, insert the *3–D Grapher* disk. Otherwise type *3DG.EXE* (or *3DG-H.EXE* if you have a Hercules graphics card). Once you start the program, information about the graphing program will appear on the screen. Press any key and a graph will be drawn.

## 11.9  3-D Grapher Menu

**11.9.1  Aiming Point:**  Select Ⓐ to choose a new point at which the viewing eye will be focused. You will need to enter three parameters, $x$, $y$, and $z$, which represent the rectangular coordinates of the desired aiming point.

**11.9.2  Clear Screen:**  Select Ⓒ to clear the current graphing window.

**11.9.3  Set Cuts (Draw):**  Select Ⓓ to choose the axis through which the points will be plotted. Default is " $x$-cuts" where sections are taken parallel to the $yz$ plane.

**11.9.4  Elevate:**  Select Ⓔ to change the elevation or the angle $\phi$ of the viewing point. $\phi$ is the angle the line through the origin and the viewing point makes with the $z$-axis. The default elevation angle $\phi$ is $60°$ .

**11.9.5   Change F($x, y$):** Press ⟨F⟩ to change the function to be graphed. You will see a menu with 10 options. Pressing ⟨1⟩ through ⟨8⟩ will select the function to be plotted. Only one *Selected* function can be plotted. Thus, you can *predefine* up to 8 functions and then plot them individually. Press ⟨9⟩ to return to the default menu.

Select ⟨0⟩ to change any one of the eight functions listed. When you press ⟨0⟩, you will then be asked which function index you wish to change. (There will be no prompt.) You will now be able to edit the function you specified; to do this use the keys in the table below. Edit the equation you will want to graph (in terms of $x$ and $y$) and press ⟨return⟩.

See Section 11.2.4 for a list of the editing keys, the special symbols, and the built-in functions.

**11.9.6   Redraw Graph:** Select ⟨G⟩ to replot the surface.

**11.9.7   Hidden Lines:** Select ⟨H⟩ to add or delete the hidden line subroutine. This command is basically a toggle switch between *Hidden Lines* and *No Hidden Lines*.

**11.9.8   Add Axis:** Select ⟨I⟩ to add the three coordinate axes to the graph.

**11.9.9   Lines/Grid:** Selecting ⟨L⟩ will bring up two menus, a "Line Menu" and a "Grid Menu", with six options under the line menu and three under the grid menu. The line menu will allow you to graph a line in 3-D space, where the user will supply the desired plane and the position on the plane. After this is done the line will be drawn from the range of the viewing cube plus four. For example, if you wanted to draw a line from (1,2,zmin−2) to (1,2,zmax+2) you would do the following: press ⟨L⟩ to enter the "Lines/Grid" menu, press ⟨4⟩ to draw a line "IN $xz$ PLANE AT $x$", enter the $x$ parameter by pressing ⟨1⟩⟨return⟩, enter the $y$ parameter by pressing ⟨2⟩⟨return⟩. The line will now be drawn given the parameters. The "Grid Menu" has three options. Simply select the plane you want a grid drawn in by selecting one of the three options. At this point you will be prompted for the plane position. Type any real number or algebraic expression and the grid will appear at that position in the desired plane.

**11.9.10   Rotate:** Select ⟨R⟩ to change the rotation angle $\theta$. $\theta$ is the angle between the $x$-axis and the plane perpendicular to the $xy$-axis which contains the viewing point. The default rotation angle $\theta$ is 30°.

**11.9.11   Resolution (Speed):** Select ⟨S⟩ to change the resolution (that is, the number of sections and number of points per section). When you select ⟨S⟩ you will be asked to enter two parameters, the "Number of Sections", and the "Number of Points/Section". The number of sections refers to the number of $x$-sections ($x$-cuts or $x$-wires), that is, sections *parallel* to the $yz$ plane. A given $x$-section, say $x = a$, is the graph of $z = f(a, y)$ (the intersection of the graph of the surface and the plane $x = a$). The larger the number, the slower the plotting speed, but you get better resolution with more sections. The number of points/section refers to the number of points that are to be drawn in one $x$-section, or $yz$ plane. Again, more points means slower plotting speed but better resolution.

**11.9.12   Defaults:** Select ⟨T⟩ to return to the default mode for all features.

**11.9.13   Change Viewing Box:** Select ⟨V⟩ to change the viewing box. When you select ⟨V⟩ you will see the old viewing box displayed and then you will be prompted to input *xmin, xmax, ymin, ymax, zmin,* *zmax* (in that order). Think of *xmin* as "back" and *ymin* as "left", etc. (see Figure 11.8). If you want the new viewing box to be [−20, 30] by [−15, 40] by [−10, 10] you will enter ⌐-⌐⌐2⌐⌐0⌐ ⌐return⌐ ⌐3⌐⌐0⌐ ⌐return⌐ ⌐-⌐ ⌐1⌐ ⌐5⌐ ⌐return⌐ ⌐4⌐⌐0⌐ ⌐return⌐⌐-⌐ ⌐1⌐ ⌐0⌐ ⌐return⌐ ⌐1⌐⌐0⌐ ⌐return⌐. The new graph will immediately be calculated and then drawn.

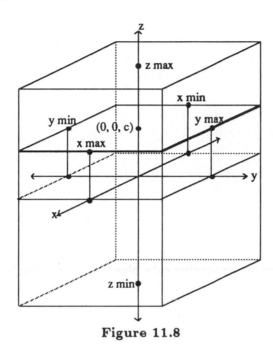

**Figure 11.8**

**11.9.14  Exit:** Select ⌐X⌐ to exit the program and return to the IBM $^R$ Operating System.

**11.9.15  Change Distance (Zoom):** Select ⌐Z⌐ to change the *distance* from the viewing point to the aiming point. A distance of 100 causes the viewing box [−10, 10] by [−10, 10] by [−10, 10] to approximately "fill" the computer screen. The default zoom distance is 100.

## 11.10  Special Features for the 3-D Grapher

This section will discuss the special features available to the user.

### 11.10.1  Special Keys

(1)  **Abort Plot.** To STOP the plot at any time during the plotting or computation of the graph, press the ⌐Esc⌐ key or ⌐E⌐ key.

(2)  **Pause Plot.** To interrupt (PAUSE) the plot, press the ⌐spacebar⌐. Pressing the spacebar again or any other key will RESUME the plot. The $x$-, $y$-, and $z$-coordinates of the last point plotted (in the current section being plotted) will be displayed.

### 11.10.2  Helpful Hints

(1)  Do not turn the hidden line option on (key $\boxed{\text{H}}$) until you finish your exploration because the plot will be slower with the hidden line option active.

(2)  Because of the increased plotting time, increase resolution only when you desire a "nice" plot.

(3)  To plot a *sphere*, for example, $z^2 + y^2 + x^2 = 70$ (and other similar quadratic surfaces), first plot $z = (70 - x^2 - y^2)^{(1/2)}$. Then select $\boxed{\text{F}}$ to change the "displayed" function to $z = -(70 - x^2 - y^2)^{(1/2)}$. Finally, click $\boxed{\texttt{previous menu}}$ and the "lower" hemisphere will be overlayed on the existing "upper" hemisphere.

(4)  Use the zoom-in key $\boxed{\text{Z}}$ to change the cone of vision. Changing the distance $d$ allows you to see more, or less, of the graph that is contained in the viewing box.

(5)  To zoom-in on features away from the center (aiming point) of a current graph, you may need to change both the viewing box *and* the aiming point. However, do this in low resolution until you are sure you have selected the viewing box and the aiming point that displays the desired features.

---

*Note*   **AT&T users**

If you are using an AT&T series computer with an 8087 math coprocessor you will need to disable the math coprocessor for the software to run. This can be accomplished by typing "set no87=suppressed $\boxed{\texttt{return}}$". To enable the 8087 after the program is finished, simply type "set no87= $\boxed{\texttt{return}}$".

# Chapter 12

## Master Grapher Version 1.0 for the Macintosh™

### 12.1 *Master Grapher* Start Up

**IMPORTANT.** Do NOT write protect either the *MASTER GRAPHER* or *3-D GRAPHER* disk; the program writes to the disk.

**Start Up.** Boot with a system disk (use system 4.1 and finder 5.5 or better), then insert the *Master Grapher* disk. If you are not using Finder 5.5 or higher or are using a Mac 512 you will need to double click on the *Master Grapher 512* icon, otherwise double click on the *Master Grapher* icon. To determine which system and finder you are using pull down the Apple menu and get About Finder while in the finder. A dialog box will appear with the finder number on it. Once you start the program information about the graphing program will appear on the screen. Click the mouse or press any key to go to the *main menu*.

**3-D Grapher.** *Master Grapher* includes *3-D Grapher*, a graphing utility for functions of *two* variables. The *3-D Grapher* utility is described in Sections 12.8–12.12.

### 12.2 Modifying the Program Defaults

Before running any of the graphing programs on this disk you may choose to modify some of the program defaults. When you press ⑦ or click the mouse on *7. Modify Program Defaults* you will see Figure 12.1.

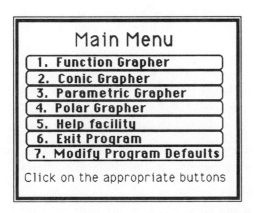

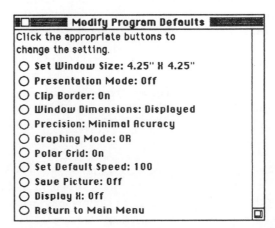

Figure 12.1

**12.2.1  Set Window Size:**  The first option on the list, *Set Window Size*, can be used to set the size of the graphing window. When you click the *Set Window Size* button, the screen will clear and the current window size and the maximum window size will be displayed. The program will then prompt you to enter the width and then the height of the window (in terms of inches). There are advantages to a square screen, but you can choose any size window provided it fits within a minimum of 0.75 inches and a maximum dimension which is determined by the machine you are using. If you enter a dimension which is too small or too large the program will choose the maximum viewing size for that dimension.

**12.2.2  Presentation Mode:**  The next option, *Presentation Mode* allows the user to make the grapher more visible if the user is displaying it on an overhead projector (in *On* position). This is achieved in the program by using thicker lines to draw the graph and using larger and boldface text. The thicker lines are not as clear on the screen, so this option should only be used for presentation purposes.

**12.2.3  Clip Border:**  The next option, *Clip Border* is used to toggle between *Clip Border: On* and *Clip Border: Off.* When you use the option *Clip Graphics Screen*, the graph is copied to the clipboard and can be carried into the scrapbook. The clip border option lets you decide whether it will be copied with or without the border.

**12.2.4  Window Dimensions:**  *Window Dimensions: Displayed* gives you the option of having the L, R, B, and T values displayed in the viewing rectangle or having the *Window Dimensions: Not Displayed* so that the screen is less cluttered. If these parameters are not displayed, the equation of the function being graphed is also not displayed. However, if either dimension of the window is less than or equal to 2.5 inches the dimensions are not displayed in the window but the function is displayed in another window. When you use *Master Grapher* to draw a graph to be used for demonstration, you may prefer to use *Window Dimensions: Not Displayed*. However, when you are using *Master Grapher* to explore and discover the behavior of a relation, particularly when zooming in or out, you will most likely choose to use *Window Dimensions: Displayed*.

**12.2.5  Precision:**  When you use *Window Dimensions Not Displayed*, notice that the next option becomes shadowed. This option, *Display Minimal Accuracy*, is available only with *Window Dimensions Displayed*, because it determines the accuracy of the numbers used to represent the L, R, B, and T values. *Precision: Minimal Accuracy* toggles between its default setting and *Precision: Full Precision*, and determines whether the L, R, B, and T values are given with full machine precision accuracy or with significant digit accuracy.

**12.2.6  Graphing Mode:**  The *Graphing Mode: OR* is a toggle between *Graphing Mode: OR* and *Graphing Mode: XOR* that determines whether the intersection of two lines is represented as an open pixel (XOR, exclusive or) or with merely a crossing of lines with no change in the pixels at the intersection (OR, or).

**12.2.7  Polar Grid:**  Setting *Polar Grid: On* will cause a polar grid to be used in the polar grapher instead of the rectangular one that is used in all the other graphers. If *Polar Grid: Off* is set then a rectangular grid will be drawn in the polar grapher.

**12.2.8  Default Speed:**  The *Set Default Speed* option will allow you to modify the program's default speed. This is not a direct change to the default speed but a rule change, where the following rule is applied.

| Rule 1 | - Function grapher: | Factor * 1 * Window Size/4.45 |
| Rule 2 | - Conic & Parametric grapher: | Factor * 0.5 * Window Size/4.45 |
| Rule 3 | - Polar grapher: | Factor * 2 * Window Size/4.45 |

where Window Size is the width of the window. For example, to obtain a default speed of 100 in the conic grapher given the width of the window to be 4.45 inches, the default speed setting must be 200.

**12.2.9  Save Picture:** The *Save Picture* option saves the graphing window to disk when the user clicks on any of the buttons in the grapher menu. This option can be useful to instructors, but it slows the program down and can use quite a bit of disk space. It is recommended that you have a hard drive when you use this option because each file uses a minimum of 1 K of disk space, but on the average it uses 4K of disk space. The files are saved in Microsoft $^{TM}$ Basic graphic file format so you can write a basic program to view them.

**12.2.10  Display X:** The next option, *Display X*, will allow the user to watch the values of $x$ or $t$ as the graph is being drawn. Although this is quite a useful item, it is recommended that you leave it off unless you have an accelerated Mac + or SE or you have a Mac II or better because of speed considerations.

**12.2.11  Return to Main Menu:** This last option will return you to the main menu.

## 12.3  Function Grapher

When you are ready to enter the Function Grapher, select $\boxed{1}$ from the Main Menu. The initial screen (except for the function graphed and this viewing rectangle) will look like the one shown in Figure 12.2.

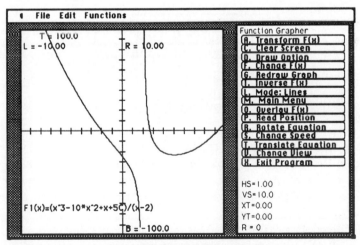

**Figure 12.2**

The available interactive commands will be in the window on the right side of the screen. In the bottom of this window are listed the distance between horizontal scale marks (HS) and the distance between the vertical scale marks (VS), as well as the translation factors (XT and YT), and the rotation factors (R).

**12.3.1  Transform F(x):** This command allows you to draw the graph of $y = A f (B x + C) + D$ by specifying $f$, A, B, C, and D. Press $\boxed{A}$ or click the mouse on *A. Transform F(x)*. You will then be asked to enter the index of the function $f$ you wish to transform. For example, if you choose 6, then the sixth function in the function menu will be the function you are going to transform. Next, four edit fields and an *OK* button will appear. Enter the parameters A, B, C, and D. You may enter any real number or algebraic expression, for example "pi*2". To move to a different field either press the [tab] key or click the mouse on the other edit fields you want to modify. When all the parameters have been changed to the desired values press the [return] key or click the *OK* button. The function $y = A f (B x + C) + D$ will be plotted immediately.

**12.3.2  Clear Screen:**  Select [C] to clear the graphing window.

**12.3.3  Draw Option:**  When you select [D], a menu like the one below will appear with a list of options. Use [1] through [4] to approximate the coordinates of a point on the graph. To approximate the $y$-coordinate, a horizontal line can be specified to move up and down (use the [↑] or [U] to move the line up, the [↓] or [D] to move the line down, and [S] to change the rate of movement of the line) until it is fixed in position by pressing any other key. The $y$-coordinate of each point on the horizontal line will be displayed. The $x$-coordinate can be found in a like manner using a vertical line (use the [→] or [R] and the [←] or [L] to move it right and left). Select [5] to overlay a lattice (an array of dots) in the viewing rectangle. The *distance* between these dots will be given by the HS (horizontal scale) and VS (vertical scale) values found in the lower right-hand corner of the screen. This command will help estimate the coordinates of a point on a graph. For example, zoom-in on some area of the current graph that is of interest to you. Select [D] [5] to obtain a lattice in the viewing rectangle. Use the HS and VS values to read the coordinates of the point you selected.

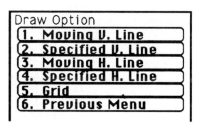

**12.3.4  Change F(x):**  Select [F] to change the function to be graphed. You will see a menu like the one in Figure 12.3. Clicking on the buttons **F1(x)=** through **F8(x)=** will either select or deselect them. Only the *Selected* function(s) will be plotted. Thus, you can *predefine* and plot up to 8 functions at the same time. Clicking on the *Previous Menu* button returns you to the default menu.

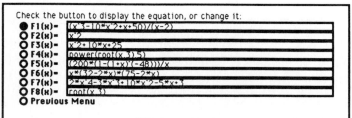

**Figure 12.3**

To edit an equation, click the mouse on the boxed equation you want to change. You will now see either an area of the equation highlighted or a text cursor. To change the whole equation simply use the mouse to highlight all of the text and then start typing the new equation. The *Cut, Copy, Paste* option under the edit menu will function properly while editing the equation. Make note the $\boxed{\texttt{tab}}$ does not move you from field to field, and the $\boxed{\texttt{return}}$ does not enter the equation.

Use standard BASIC syntax to enter the desired equation, for example $x^4 - 3x^2 + 15$ would be entered by pressing $\boxed{\texttt{x}}$ $\boxed{\texttt{$\wedge$}}$ $\boxed{\texttt{4}}$ $\boxed{\texttt{$-$}}$ $\boxed{\texttt{3}}$ $\boxed{\texttt{$*$}}$ $\boxed{\texttt{x}}$ $\boxed{\texttt{$\wedge$}}$ $\boxed{\texttt{2}}$ $\boxed{\texttt{$+$}}$ $\boxed{\texttt{1}}$ $\boxed{\texttt{5}}$. The constants $e$ and $\pi$ are entered as $\boxed{\texttt{e}}$ and $\boxed{\texttt{p}}\boxed{\texttt{i}}$, respectively. Special keying instructions are needed to enter built-in functions or to obtain correct graphs of some special functions. Here is a list of the special functions.

## Special Symbols and Built-in Functions

( 1) $+$ is addition.

( 2) $-$ is subtraction.

( 3) $*$ is multiplication.

( 4) $/$ is division.

( 5) $\wedge$ is $x^a$.

( 6) $\backslash$ is Integer Division.

( 7) ABS$(x - 2)$ is $|x - 2|$.

( 8) CEIL$(x) = [x + 1]$.

( 9) EXP$(x)$ is $e^x$.

(10) FIX$(x)$ is FLOOR if $x > 0$, and CEIL if $x < 0$.

(11) FLOOR$(x) = [\![x]\!]$.

(12) INT$(x)$ is the greatest integer function .

(13) LOG$(x)$ is $\ln x$.

(14) LOG10$(x)$ is $\log_{10}(x)$.

(15) LOG2$(x)$ is $\log_2(x)$.

(16) ROUND$(x)$ rounds to the nearest integer.

(17) SGN$(x)$ is the signum function.

(18) SQR$(x + 6)$ is $\sqrt{x + 6}$.

(19) SIN$(x)$ is sin $x$. All other trigonometric functions are entered in the same manner (e.g. arctan $x$ is ARCTAN$(x)$, cosh $x$ is COSH$(x)$, etc.). Here is a list of all the trigonometric functions supported: arccos, arccosh, arccot, arccoth, arccsc, arccsch, arcsec, arcsech, arcsin, arcsinh, arctan, arctanh, cos, cosh, cot, coth, csc, csch, sec, sech, sin, sinh, tan, tanh.

## Special Functions

(20) LOGB$(x, a)$ is $\log_a(x)$.

(21) ROOT$(x, a)$ is $x^{1/a}$. Note: You can enter $x^{1/a}$ as $x \wedge (1/a)$, but you will only get the portion of the graph in the first quadrant.

(22) POWER$(x, a)$ is $x^a$.

**12.3.5  Redraw Graph:** Redraws the graph with the current settings.

**12.3.6  Inverse F(x):** If you wish to view the inverse relation $(y, x)$ where $y = f(x)$ of any function in the function menu, select $\boxed{\texttt{I}}$. Then select the index of the function you wish to invert and press $\boxed{\texttt{return}}$, and the inverse *relation* will be overlayed immediately.

**12.3.7  Mode: Lines or Points:** Select $\boxed{\texttt{L}}$ to choose one of two plotting modes. One plots only the points evaluated for a particular function, whereas the second connects each consecutive pair of points with a line segment (the default mode).

**12.3.8  Main Menu:** Select $\boxed{\texttt{M}}$ to return to the main menu.

**12.3.9  Overlay F(x):**  When you select $\boxed{\text{O}}$ you will need to select an index function to overlay and then press $\boxed{\texttt{return}}$.  For example, if you choose 6, then the sixth function in the function menu will be plotted on the same screen with the function(s) already plotted.  Once the index is selected you will need to tell the program whether you want the function rotated and translated or not.  If there are non-zero values in the XT, YT, and R categories in the lower right hand corner of the display screen, then choosing $\boxed{\text{1}}$ and pressing $\boxed{\texttt{return}}$ will plot the selected function *with* those rotations and translations applied.  Choosing $\boxed{\text{0}}$ will have the function plotted *without* any rotation or translation applied.

**12.3.10  Read Position:**  The next option, $\boxed{\text{P}}$, can also be used to approximate the coordinates of any point in the current viewing rectangle.  After selecting $\boxed{\text{P}}$ simply click the mouse on the desired position in the graphing window.  In the lower right-hand corner you will now see the location of the cursor in the viewing rectangle.  To read another position, simply follow the procedure again.

**12.3.11  Rotate F(x):**  Select $\boxed{\text{R}}$ to rotate the function about the origin.  When you select $\boxed{\text{R}}$, you will need to enter the counterclockwise rotation angle in degrees.  This can be either a real number or an algebraic expression like "pi*2".  A positive input will produce a counterclockwise rotation and a negative input will be clockwise.

**12.3.12  Change Speed:**  To change the plotting speed, select $\boxed{\text{S}}$.  You will now see the current plotting speed, the minimum and maximum allowable plotting speeds, and a prompt to enter a new speed.  Enter the speed you want by typing either a number or an algebraic expression with terms pi, e, or some constants and then press $\boxed{\texttt{return}}$.  The program will evaluate the expression and enter the resulting value as the new plotting speed.  (You may wish to experiment with various settings until you find the one you prefer.)  The default speed is a good compromise between speed and resolution.

**12.3.13  Translate F(x):**  Select $\boxed{\text{T}}$ to draw a graph of the function $y = f(x)$ translated H units horizontally and V units vertically.  After selecting $\boxed{\text{T}}$, enter the amount of horizontal translation and press $\boxed{\texttt{return}}$.  Then enter the amount of vertical translation and press $\boxed{\texttt{return}}$.  Nothing will appear to happen.  To see the effect of the translation, you will need to redraw the graph $\boxed{\text{G}}$.  You will now have graphed the *displayed* function with the translation factors applied. If you wish to view both the original function and the translated function, select $\boxed{\text{O}}$ (Overlay F(x)) and choose the index of the original function and then $\boxed{\text{0}}$ W/O Rot. & Trans. In this manner you can get both the function and its translation in the same viewing rectangle.

**12.3.14  Change View:**  Selecting $\boxed{\text{V}}$ will display the menu on the next page.  You may have changed the speed, the plot mode or a variety of different commands that do not cause the graph to be immediately replotted.  The function grapher incorporates those changes when the graph is redrawn or the viewing rectangle is changed.  Suppose we want to view the graph of

$$f(x) = \frac{x^3 - 10x^2 + x + 50}{x - 2}$$

in the viewing rectangle $[-1, 1]$ by $[-1, 1]$.  To do this enter $f$ by selecting $\boxed{\text{F}}$ and proceeding as detailed in the "Change F(x)" section (12.3.4), then select $\boxed{V}$. You can now choose the method you want to change the viewing rectangle.

View Menu
0. Zoom In
1. Zoom In (Point)
2. Zoom Out
3. Zoom Out (Point)
4. Set Zoom Factors
5. Set Window
6. Default Window
7. Last View
8. H-Scale: 1
9. Previous Menu

(0) **Zoom-In.** To select a new viewing rectangle by drawing it directly on the screen, select $\boxed{0}$. Move the mouse to the desired upper left hand corner of the new viewing rectangle within the current view. Now press the mouse button and, while holding it down, drag the mouse to the lower right hand corner of the viewing rectangle. When you release the mouse button you will now see the function plotted in the new viewing rectangle.

An illustration of the effect that "zoom-in" can have on a given view appears in Figure 12.4. The function $f(x) = x^2 - 3x + 2$ was graphed in the standard viewing rectangle, then graphed again in the $[0.8, 2.3]$ by $[-1, 0.5]$ viewing rectangle. The behavior of the graph between $x = 1$ and $x = 2$ is much more obvious in the zoom-in view.

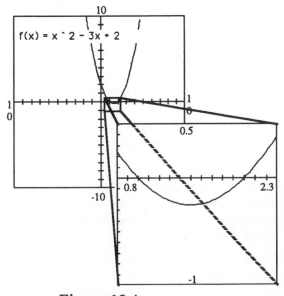

**Figure 12.4**

(1) **Zoom In (Point).** This command is useful when you want a close-up or magnified view of a particular section of a graph around a particular point. Select $\boxed{1}$. Now move the mouse to the point you want to be the center of the new viewing rectangle and click the mouse. The function will be replotted in the new viewing rectangle with the point you selected in its center. The *size* of the viewing rectangle will be determined by the zoom factor settings $\boxed{4}$.

(2) **Zoom Out.** When you select $\boxed{2}$, the graph will be immediately redrawn in a viewing rectangle expanded by the zoom factors you set with key $\boxed{4}$ (or the default values of 10 for $x$ and 10 for $y$).

(3) **Zoom Out (Point).** When you select $\boxed{3}$, you will be asked to click the mouse in the current viewing rectangle at the location about which you wish the "zoom-out" to be centered. Once again the zoom factor will be determined by the default factors (both 10) or the factors you set using $\boxed{4}$.

(4) **Set Zoom Factors.** This command sets the horizontal ($x$) and vertical ($y$) zoom factors. When you select $\boxed{4}$ you will see the old settings for $x$ and $y$ zoom factors and then you will be prompted to enter new ones. This can be done by typing in any real number or algebraic expression. Assume the current viewing rectangle is [L, R] by [B, T]. Enter the value you wish [L, R] and [B, T] to be multiplied by. Values greater than 1 cause the horizontal or vertical size of the rectangle to increase (zoom-out), and values less than 1 cause the horizontal or vertical size of the rectangle to decrease (zoom-in). The change in size of the viewing rectangle is symmetric about the center of the rectangle. Sometimes you will want the zoom factors on the $x$-axis and $y$-axis to be different. Note: You can "zoom-in" using "zoom-out" by selecting zoom factors less than one.

(5) **Set Window.** To change the viewing rectangle, select $\boxed{5}$. The menu will clear and an *OK* button and four edit fields labelled *L, R, T, B* will appear. Change the values using the edit menu, the mouse, and the keyboard. You can move to a different field by either pressing $\boxed{\text{tab}}$, or click on the other field. When you are done editing the values simply press $\boxed{\text{return}}$, or click the *OK* button. The values can be any real number or algebraic expression. The screen will be redrawn in the specified viewing rectangle and the "View Menu" will return.

(6) **Default Window.** When you select $\boxed{6}$ the displayed function(s) will be replotted in the default viewing rectangle $[-10, 10]$ by $[-10, 10]$, default speed, mode: lines, rotation $0°$, and translation $(1, 1)$. After the screen is redrawn you will be returned to the "View Menu".

(7) **Last Window.** When you select $\boxed{7}$ the displayed function(s) will be replotted in the last viewing rectangle. However, speed, mode, rotation, and translation will remain unaltered from the current settings. After the screen is redrawn you will be returned to the "View Menu".

(8) **X-Scale.** When you select $\boxed{8}$, the distance between the tick marks on the $x$-axis will be changed. If X-Scale is in units of $\frac{\pi}{2}$ the distance between tick marks will be in a factor of units of either $\frac{\pi}{2}$ or $\frac{\pi}{4}$. If X-Scale is in units of 1 the distance between tick marks will be in a factor of units of either 1 or 0.5. The graph will be redrawn immediately with the appropriate $x$-scale. This option is available only in the function grapher.

(9) **Previous Menu.** When you select $\boxed{9}$, you will return to the grapher menu. This option is $\boxed{8}$ on all the other graphers.

**12.3.15 Exit:** Selecting $\boxed{\text{X}}$ will exit the program and return you to the system.

## 12.4  Conic Grapher

When you are ready to enter the Conic Grapher, select $\boxed{2}$ from the Main Menu. There will be a slight delay while the conic grapher program is loaded. The initial screen (except for the conic equation graphed) will look like the one shown in Figure 12.5.

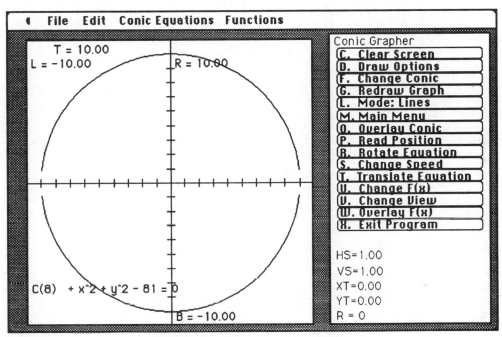

**Figure 12.5**

Notice the changes in the conic grapher menu versus the function grapher menu. The "A. Transform F(x)" and "I. Inverse F(x)" options have been removed. The "F. Change F(x)" is now "F. Change Conic" and the "O. Overlay F(x)" is now "O. Overlay Conic". Notice the "F(x)" in rotate and translate has been changed to equation. Two new options have been added: "U. Change Function(s)" and "W. Overlay Function(s)". In the following sections we will only discuss these features as all the others have not changed. Note: Rotate and translate have not been changed, only the wording in the menu has been changed.

**12.4.1  Change Conic:** Pressing $\boxed{F}$ will clear the viewing rectangle and you will see a list of eight *conic equations*. Clicking on any of the C(1)–C(8) buttons will allow you to toggle between *Displayed* and *Not Displayed*, where an equation is displayed if it is selected. Clicking on *Previous Menu* will take you back to the conic grapher menu. When you click on *Change Conic Equation* you will see seven edit fields. The index of the function is the most important field. It must be set first because it determines the values of all the other parameters. The next six fields are the parameters A, B, C, D, E, and F. These parameters are the constants in the following equation: $Ax^2 + Bxy + Cy^2 + Dx + Ey + F = 0$. The constants may be entered as real numbers or as algebraic expressions. Once the parameters are set, either press $\boxed{\texttt{return}}$ or click the *OK* button. You will then see the changes appear in the menu.

Check the button to display the equation, or check change:

O **Change Conic Equation**
● C(1) + x^2 + y^2 - 81 = 0
O C(2) - x + 3y - 5 = 0
O C(3) + 2x^2 - y^2 + 3x - 5 = 0
O C(4) - x^2 + 2y^2 + 3y - 5 = 0
O C(5) + xy - x + 3y - 5 = 0
O C(6) + x^2 + 2xy + y^2 + 2x + 2y - 81 = 0
O C(7) + x^2 + 3xy + x + y = 0
O C(8) - 2x^2 + 3y - 5 = 0
O **Previous Menu**

**Figure 12.6**

**12.4.2 Overlay Conic:** Selecting ⊡ will allow you to overlay a *conic equation* in the same manner as you did with functions in Section 12.3.9.

**12.4.3 Change Function(s):** Selecting ⊡ will allow you to change or display any of the eight *functions* in the same manner as you did in Section 12.3.4.

**12.4.4 Overlay Function:** Selecting ⊡ will allow you to overlay one of the eight *functions* in the same manner as you did in Section 12.3.9.

## 12.5 Parametric Grapher

When you are ready to enter the parametric grapher, select ⊡ from the Main Menu. There will be a slight delay while the parametric grapher program is loaded. The initial screen (except for the parametric equation graphed) will look like the one shown in Figure 12.7.

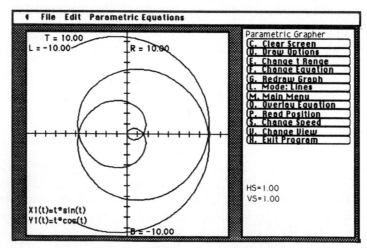

**Figure 12.7**

Notice the changes in the parametric grapher menu versus the function grapher menu. The "Transform F(x)", "Inverse F(x)", "Rotate Equation" , and "Translate Eqn." options have been removed. The

"F.  Change F(x)" is now "F.  Change Parametric" and the "O.  Overlay F(x)" is now "O.  Overlay Parametric".  One new option has been added: "E.  Change t Range".  In the following sections we will only discuss these new features as all the others have not changed.

**12.5.1  Change Parametric:**  Select $\boxed{\text{F}}$ to change the parametric equation to be graphed.  You will see a menu with five buttons and eight edit fields.  Selecting and deselecting $\boxed{\text{1}}$ through $\boxed{\text{4}}$ will determine which equation sets are to be graphed, the same way as you did in Section 12.3.4.  Thus, you can *predefine* and plot up to 4 parametric equations in this manner.  Select $\boxed{\text{5}}$ to return to the default menu.

   To change any one of the four parametric equations listed, simply edit them in the same manner as you did in Section 12.3.4.  Remember that these are parametric equations so you need to edit the equation pair and use $t$ as the variable.

**12.5.2  Overlay Parametric:**  Selecting $\boxed{\text{O}}$ will allow you to overlay a *parametric equation* in the same manner as you did in Section 12.3.9.  Note: There are only four indices, not eight as in all the other graphers.

**12.5.3  Change t Range:**  Select $\boxed{\text{E}}$ to change the bounds of $t$.  The default bounds on the parameter $t$ are $-10 < t < 10$.  After selecting $\boxed{\text{E}}$ you will see the current range of $t$ and then you will be prompted to enter tmin and tmax.  These parameters can be any real numbers or algebraic expressions like "pi*2".

## 12.6  Polar Grapher

When you are ready to enter the polar grapher, select $\boxed{\text{4}}$ from the Main Menu.  There will be a slight delay while the polar grapher program is loaded.  The initial screen (except for the polar equation graphed) will look like the one shown in Figure 12.8.

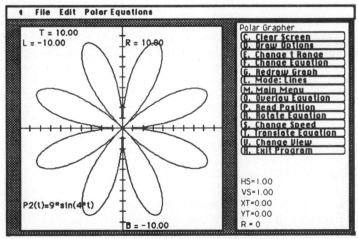

**Figure 12.8**

   Notice the changes in the polar grapher menu versus the function grapher menu.  The "A.  Transform F(x)" and "I.  Inverse F(x)" options have been removed.  The "F.  Change F(x)" is now "F.  Change polar" and the "O.  Overlay F(x)" is now "O.  Overlay Polar".  Notice the "F(x)" in rotate and translate has been changed to "Equation". Finally, notice the new feature "E. Change t Range". In the following sections we will only discuss these features as all the others have not changed. Note that rotate and translate have not been changed, only the wording in the menu has been changed.

**12.6.1  Change Polar:** Selecting $\boxed{F}$ will allow you to change or display any of the eight *polar equations* in the same manner you did in Section 12.3.4. Remember to use $t$ as the variable, not $x$.

**12.6.2  Overlay Polar:** Selecting $\boxed{O}$ will allow you to overlay a *polar equation* in the same manner you did in Section 12.3.9.

**12.6.3  Change t Range:** Selecting $\boxed{E}$ will allow you to change the t range in the same manner you did in Section 12.5.3.

## 12.7  Special Features for *Master Grapher*

This section will discuss the special features available to the user such as the menus and special keys.

### 12.7.1  File Menu

(1)  **Quit.**  This option allows you to quit the program from almost anywhere. The exception to this rule is if you are entering a number in the program, you will be able to quit only after the number is entered.

(2)  **Print Graphics Screen.**  This item will print the graphics window with or without a border (see Section 12.2.3) to the available printer. The available printer is set by chooser.

(3)  **Clip Graphics Screen.**  This item will clip the graphics window with or without a border. (See Section 12.2.3 to set the clip border.) To use this option, select this menu item and then open up the desk accessory "Scrapbook". Paste the clipboard into the Scrapbook and then close the Scrapbook. It is important to remember to close the Scrapbook. If it is left open, this program cannot clip to the clipboard. (In some paint programs like "SuperPaint $^{TM}$" this is also true.)

### 12.7.2  Edit Menu

(1)  **Undo.**  This option is not available.

(2)  **Cut.**  This allows you to cut the text out of any edit field.

(3)  **Copy.**  This allows you to copy the text from an edit field to the clipboard.

(4)  **Paste.**  This allows you to paste any text in the clipboard to an edit field.

**12.7.3  Function Menu, Conic Equations, Parametric Equations, Polar Equations:** All of these menus do basically the same thing: they allow you to select (or deselect) the equations that are going to be graphed when you select $\boxed{G}$ *Redraw Graph*.

### 12.7.4  Special Keys

(1)  **Abort Plot.**  To STOP the plot at any time during the plotting routine and return to the menu, press the $\boxed{\text{Esc}}$ key or $\boxed{E}$ key.

(2) **Pause Plot.** To interrupt (PAUSE) the plot, press the $\boxed{\texttt{spacebar}}$. Pressing the spacebar again or any other key will RESUME the plot.

## 12.8  3-D Grapher Introduction

A great deal is known about single variable function graphers. The use of such graphers and understanding about how they can enhance the teaching and learning of mathematics are on the rise. Much less is known about the use of surface graphers, that is, devices that produce a graph of a function of two variables. However, several things about graphing functions of two variables are very clear. Obtaining graphs by hand is a difficult task for both student and teacher. Students have a good bit of trouble visualizing in three dimensions. Teachers have a rough time producing quick, accurate graphs of functions of two variables.

The three-dimensional grapher described in this manual is designed to allow the user to obtain reasonably accurate graphs of functions of two variables. The user can obtain the graph for $a \leq x \leq b$, $c \leq y \leq d$ and $e \leq z \leq f$, and then choose an arbitrary point in three dimensional space from which to view the graph. Once the first graph is drawn, the points are stored in an array so that the graph can be redrawn quickly from different views. The user can choose any point in three-dimensional space from which to view the graph. The resolution of a graph is under user control.

This three-dimensional grapher allows the user to interactively explore the behavior of surfaces. Local maximum and minimum values of functions of two variables can be investigated graphically. The grapher can help students deepen understanding and intuition about functions of two variables. It can provide a geometric representation of problem situations to go along with an algebraic representation. The connections between these two representations can be explored and exploited to gain better understanding about problem situations.

The single most important feature of this graphing program is that virtually every aspect of this utility is interactive and under user control. This utility was designed to help teachers teach and students learn mathematics in an atmosphere where both are active partners in the educational experience.

We will now describe how the user chooses a region of three-dimensional space in which to draw a graph of a function of two variables, and the way in which that graph can be viewed.

*Definition* The set $\{(x, y, z) \mid a \leq x \leq b, \; c \leq y \leq d, \; e \leq z \leq f\}$ is called the *viewing box* $[a, b]$ by $[c, d]$ by $[e, f]$.

Notice that the viewing box $[a, b]$ by $[c, d]$ by $[e, f]$ is completely determined by the points $a \leq x \leq b$, $c \leq y \leq d$, $e \leq z \leq f$. The user can change the viewing box by selecting key $\boxed{\text{V}}$ and entering the values of $a, b, c, d, e,$ and $f$. The default viewing box is $[-10, 10]$ by $[-10, 10]$ by $[-10, 10]$.

Next, the user decides how to view the graph contained in the selected viewing box. Two points can be selected. The point at which the user places his/her "eye" is called the *viewing point*. The point at which the view of the eye is directed is called the *aiming point*. The aiming point can be changed by selecting key $\boxed{\text{A}}$ and inputting the rectangular coordinates of the point. The viewing point can be changed by entering the spherical coordinates $(d, \phi, \theta)$ of the point using keys $\boxed{\text{Z}}$, $\boxed{\text{R}}$, and $\boxed{\text{E}}$. The "Change Elevation" key $\boxed{\text{E}}$ allows the user to select the angle $\phi$. $\phi$ is the angle the line through the origin and the viewing point makes with the $z$-axis (Figure 12.9). The "Change Rotation" key $\boxed{\text{R}}$ allows the user to select the angle $\theta$. $\theta$ is the angle between the $x$-axis and the plane perpendicular to the $xy$-plane which contains the viewing point (Figure 12.9). Positive direction is counterclockwise. Finally, the "Chg. Dist. (Zoom)" key $\boxed{\text{Z}}$ allows the user to select the distance that the viewing point is from the aiming point. The default aiming point is $(0, 0, 0)$ and the default viewing point is $(100, 60°, 30°)$.

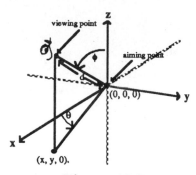

**Figure 12.9**

The software draws the graph, in a cone of vision determined by the aiming point and the viewing point. The size of the cone is an automatic feature and cannot be selected by the user. The viewing point is the vertex of the cone, and the line determined by the two points is the axis of the cone. The view is from the viewing point toward the aiming point. Selecting ⊞ allows the user to view the graph with or without hidden lines. That is, if the *hidden lines* option is on, then the user will *not* see the portions of the surface that should be hidden from view by other portions of the surface. Basically, the graph that the user sees is the intersection of the software-selected cone of vision with the user-selected viewing box. By changing the aiming point, viewing box, and viewing point, the user can view *any* portion of a surface with a high degree of resolution.

This drawing and viewing feature of the software literally allows the user to move around and view the surface as if in an airplane. You can move closer or farther away, and view the graph above or below by careful selection of elevation, rotation, and distance (Figure 12.10).

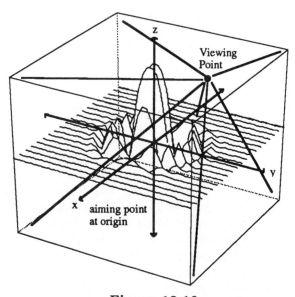

**Figure 12.10**

## 12.9  Starting Up

**Startup.** Exit the Master Grapher program by pressing $\boxed{\text{X}}$ and then do the following. If you used *Master Grapher 512* then double click on the *3-D Grapher 512* icon, otherwise double click on the *3-D Grapher* icon. To determine which system and finder you are using pull down the Apple menu and get About Finder while in the finder. A dialog box will appear with the finder number on it. If it is less than 5.5 use *3-D Grapher 512*. Once the program starts, information about the graphing program will appear on the screen. Click the mouse or press any key and a graph will be drawn.

## 12.10  3-D Grapher Menu

**12.10.1  Aiming Point:** Select $\boxed{\text{A}}$ to choose a new point at which the viewing eye will be focused. You will need to enter three parameters, $x$, $y$, and $z$, which represent the rectangular coordinates of the desired aiming point. The default aiming point is $(0, 0, 0)$.

**12.10.2  Clear Screen:** Select $\boxed{\text{C}}$ to clear the current graphing window.

**12.10.3  Set Cuts (Draw):** Select $\boxed{\text{D}}$ to choose the axis through which the points will be plotted. Default is "x-cuts" where sections are taken parallel to the $yz$ plane.

**12.10.4  Elevate:** Select $\boxed{\text{E}}$ to change the elevation or the angle $\phi$ of the viewing point. $\phi$ is the angle that the line through the origin and the viewing point makes with the $z$-axis. The default elevation angle $\phi$ is $60°$.

**12.10.5  Change F($x, y$):** Select $\boxed{\text{F}}$ to change the function to be graphed. You will see a menu like the one in Figure 12.11. Clicking on the buttons **F1(x,y)=** through **F8(x,y)=** will select the function to be plotted. Only one *Selected* function can be plotted. Thus, you can *predefine* up to 8 functions and then plot them individually. Clicking on the **Previous Menu** button will return you to the 3-D menu.

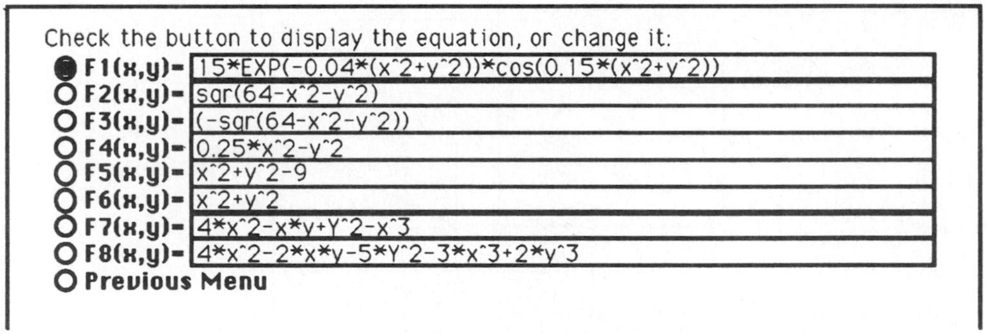

```
Check the button to display the equation, or change it:
● F1(x,y)= 15*EXP(-0.04*(x^2+y^2))*cos(0.15*(x^2+y^2))
○ F2(x,y)= sqr(64-x^2-y^2)
○ F3(x,y)= (-sqr(64-x^2-y^2))
○ F4(x,y)= 0.25*x^2-y^2
○ F5(x,y)= x^2+y^2-9
○ F6(x,y)= x^2+y^2
○ F7(x,y)= 4*x^2-x*y+Y^2-x^3
○ F8(x,y)= 4*x^2-2*x*y-5*Y^2-3*x^3+2*y^3
○ Previous Menu
```

**Figure 12.11**

To edit in this menu, click the mouse on the boxed equation you want to edit. You will now see either an area of the equation highlighted or a text cursor. To change the whole equation simply use the mouse to highlight all of the text and then start typing the new equation. The *Cut, Copy, Paste* option under the edit menu will function properly while editing the equation. Note that the $\boxed{\text{tab}}$ does not move you from field to field, and the $\boxed{\text{return}}$ does not enter the equation.

See Section 12.3.4 for the special symbols, and the built-in Functions.

**12.10.6  Redraw Graph:** Select $\boxed{\text{G}}$ to replot the surface.

**12.10.7  Hidden Lines:** Select $\boxed{\text{H}}$ to add or delete the hidden line subroutine. This command is basically a toggle switch between *Hidden Lines* and *No Hidden Lines*.

**12.10.8  Add Axis:** Select $\boxed{\text{I}}$ to add the three coordinate axes to the graph.

**12.10.9  Rotate:** Select $\boxed{\text{R}}$ to change the rotation angle $\theta$. $\theta$ is the angle between the $x$-axis and the plane perpendicular to the $xy$-axis which contains the viewing point. The default rotation angle $\theta$ is 30°.

**12.10.10  Resolution (Speed):** Select $\boxed{\text{S}}$ to change the resolution (that is, the number of sections and number of points per section). When you select $\boxed{\text{S}}$ you will be asked to enter two parameters, the "Number of Sections", and the "Number of Points/Section". The number of sections refers to the number of $x$-sections ($x$-cuts or $x$-wires), that is, sections *parallel* to the $yz$ plane. A given $x$-section, say $x = a$, is the graph of $z = f(a, y)$, the intersection of the graph of the surface and the plane $x = a$. The larger the number the slower the plotting speed, but you get better resolution with more sections. The number of points/section refers to the number of points that are to be drawn in one $x$-section, or $yz$ plane. Again, more points means slower plotting speed but better resolution.

**12.10.11  Defaults:** Select $\boxed{\text{T}}$ to return to the default mode for all features.

**12.10.12  Change Viewing Box:** Select $\boxed{\text{V}}$ to change the viewing box. When you select $\boxed{\text{V}}$ you will see the old viewing box displayed and then you will be prompted to input *xmin, xmax, ymin, ymax, zmin, zmax* (in that order). Think of *xmin* as "back" and *ymin* as "left", etc. (see Figure 12.12). If you want the new viewing box to be [−20, 30] by [−15, 40] by [−10, 10] you will enter $\boxed{-}\boxed{2}\boxed{0}$ $\boxed{\text{return}}$ $\boxed{3}\boxed{0}$ $\boxed{\text{return}}$ $\boxed{-}\boxed{1}\boxed{5}$ $\boxed{\text{return}}$ $\boxed{4}\boxed{0}$ $\boxed{\text{return}}$ $\boxed{-}\boxed{1}\boxed{0}$ $\boxed{\text{return}}$ $\boxed{1}\boxed{0}$ $\boxed{\text{return}}$. The new graph will immediately be calculated and then drawn.

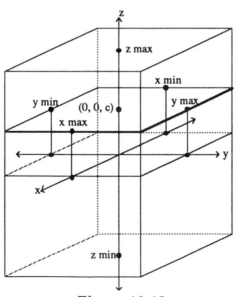

**Figure 12.12**

**12.10.13  Exit:** Select $\boxed{\text{X}}$ to exit the program and return to the Macintosh$^{TM}$ Operating System.

**12.10.14  Change Distance (Zoom):** Select $\boxed{\text{Z}}$ to change the *distance* from the viewing point to the aiming point. A distance of 100 causes the viewing box $[-10, 10]$ by $[-10, 10]$ by $[-10, 10]$ to approximately "fill" the computer screen. The default zoom distance is 100.

## 12.11  Special Features for the 3-D Grapher

This section will discuss the special features available to the user such as the menus and special keys.

### 12.11.1  File Menu

(1)  **Quit.** This option allows you to quit the program from almost anywhere. The exception to this rule is if you are entering a number in the program, you will be able to quit only after the number is entered.

(2)  **Print Graphics Screen.** This item will print the graphics window with or without a border to the available printer. The available printer is set by chooser. If you are running the *3-D Grapher 512* this option is not available.

(3)  **Clip Graphics Screen.** This item will clip the graphics window with or without a border. Here is how you should use this option. Select this menu item then open up the desk accessory "Scrapbook" now paste the clipboard into the scrapbook and then close the scrapbook. It is important to remember to close the scrapbook as if it is open this program cannot clip to the clipboard. In some paint programs like "SuperPaint$^{TM}$" this is also true. If you are running the *3-D Grapher 512* this option is not available.

### 12.11.2  Edit Menu

(1)  **Undo.** This option is not available.

(2)  **Cut.** Allows you to cut the text out of any edit field.

(3)  **Copy.** Allows you to copy the text from an edit field to the clipboard.

(4)  **Paste.** Allows you to paste any text in the clipboard to an edit field.

**12.11.3  Function Menu:** This menu allows you to select and graph any of the eight equations with the current parameters in memory (rotation, elevation, resolution, etc.).

### 12.11.4  Line Menu

(1)  **Draw line in XY plane parallel to X.** Draws a line in an $XY$ plane. The user will select the $Y$ value where the line will be drawn, and then select the $Z$ parameter to determine which $XY$ plane it is on.

(2)  **Draw line in XY plane parallel to Y.** Draws a line in an $XY$ plane. The user will select the $X$ value where the line will be drawn, and then select the $Z$ parameter to determine which $XY$ plane it is on.

(3)  **Draw line in XZ plane parallel to X.** Draws a line in an $XZ$ plane. The user will select the $Z$ value where the line will be drawn, and then select the $Y$ parameter to determine which $XZ$ plane it is on.

(4) **Draw line in XZ plane parallel to Z.** Draws a line in an $XZ$ plane. The user will select the $X$ value where the line will be drawn, and then select the $Y$ parameter to determine which $XZ$ plane it is on.

(5) **Draw line in YZ plane parallel to Y.** Draws a line in a $YZ$ plane. The user will select the $Z$ value where the line will be drawn, and then select the $X$ parameter to determine which $YZ$ plane it is on.

(6) **Draw line in YZ plane parallel to Z.** Draws a line in a $YZ$ plane. The user will select the $Y$ value where the line will be drawn, and then select the $X$ parameter to determine which $YZ$ plane it is on.

(7) **Draw Grid in XY plane.** Draws a grid in an $XY$ plane. The user will select the $Z$ parameter to determine which $XY$ plane it is on.

(8) **Draw Grid in XZ plane.** Draws a grid in an $XZ$ plane. The user will select the $Y$ parameter to determine which $XZ$ plane it is on.

(9) **Draw Grid in YZ plane.** Draws a grid in a $YZ$ plane. The user will select the $X$ parameter to determine which $YZ$ plane it is on.

## 12.11.5 Special Keys:

(1) **Abort Plot.** To STOP the plot at any time during the plotting or computation of the graph, press the $\boxed{\text{Esc}}$ key or $\boxed{\text{E}}$ key.

(2) **Pause Plot.** To interrupt (PAUSE) the plot, press the $\boxed{\text{spacebar}}$. Pressing the spacebar again or any other key will RESUME the plot. The $x$-, $y$-, and $z$-coordinates of the last point plotted (in the current section being plotted) will be displayed.

## 12.11.6 Helpful Hints:

(1) Do not turn the hidden line option on (key $\boxed{\text{H}}$) until you finish your exploration because the plot will be slower with the hidden line option active.

(2) Because of the increased plotting time, increase resolution only when you desire a "nice" plot.

(3) To plot a *sphere*, for example, $z^2 + y^2 + x^2 = 70$ (and other similar quadratic surfaces), first plot $z = (70 - x^2 - y^2)^{(1/2)}$. Then select $\boxed{\text{F}}$ to change the "displayed" function to $z = -(70 - x^2 - y^2)^{(1/2)}$. Finally, click $\boxed{\text{previous menu}}$ and the "lower" hemisphere will be overlayed on the existing "upper" hemisphere.

(4) Use the zoom-in key $\boxed{\text{Z}}$ to change the cone of vision. Changing the distance $d$ allows you to see more, or less, of the graph that is contained in the viewing box.

(5) To zoom-in on features away from the center (aiming point) of a current graph, you may need to change both the viewing box *and* the aiming point. However, do this in low resolution until you are sure you have selected the viewing box and the aiming point that displays the desired features.